T0181804

Thermal Sensors

Chandra Mohan Jha
Editor

Thermal Sensors

Principles and Applications
for Semiconductor Industries

 Springer

Editor
Chandra Mohan Jha
Intel Corporation
Chandler, AZ
USA

ISBN 978-1-4939-4963-2   ISBN 978-1-4939-2581-0   (eBook)
DOI 10.1007/978-1-4939-2581-0

Printed on acid-free paper

Springer Science+Business Media LLC New York is part of Springer Science+Business Media
(www.springer.com)

Acknowledgments

This book is an outcome of a team effort with multiple authors contributing to different chapters. I would like to thank my management in Thermal and Fluids Core Competency (ATTD, TMG) in Intel Corporation, with special thanks to Sanjoy Saha, Ashish Gupta, and our group director Gaurang Choksi. I would also like to thank our colleagues in thermal analysis team and thermal laboratory who directly or indirectly helped me complete the book.

I would like to thank all the contributing authors who have taken their personal time and effort to complete their respective chapters. The authors are Thu Huynh (DCG, Intel Corporation), Gopi Krishnan (TMG, Intel Corporation), Jaime Sanchez (TMG, Intel Corporation), Leila Choobineh and Ankur Jain (University of Texas at Arlington), S.P. Duttagupta, P. Ramesh, S. Roy, and R.A. Shukla (Indian Institute of Technology Bombay, Mumbai, India), and S.G. Kulkarni and G.J. Pathak (CMET, Pune, India).

I would like to thank Intel Press for giving me this opportunity to write a technology book, with special thanks to Patrick Hauke who helped me right from the starting phase till the publication of the book. I would also like to thank David Clark for helping me edit the book chapters. I would like to thank Springer for agreeing to publish this book. Jennifer Evans and Courtney Clark from Springer US were very helpful with their inputs about writing the technology book.

Finally, I would like to thank my family for giving me their support and allowing me to take up my personal time during weekends to write and edit this book.

This book is targeted for young engineers, scientists, researchers, and graduate students. I hope you learn the basic principles of *Thermal Sensors* and their applications in semiconductor industries.

Chandra Mohan Jha

Contents

Chapter 1
Introduction

Chandra Mohan Jha

Temperature is a key physical metric that is associated with all fields of science. It is generally linked with safety and performance and is used for important decision making in industries. In the semiconductor industry, thermal sensing and temperature measurement of a silicon device is considered a necessary step towards the qualification and certification of the device. The temperature of the operating device governs its performance and reliability. For example, a microprocessor has a certain operating temperature range and is maintained below its maximum allowed temperature for its smooth operation. An elaborate temperature control algorithm is used as a part of the thermal management of the microprocessor, where the temperature measurement of the silicon die plays an important role. The thermal management methodologies and the silicon die junction temperature evaluation procedures of the microprocessor can be applied to any active silicon device in the semiconductor industry.

There are many types of thermal sensors that exist today. Some of the commonly known thermal sensors are listed below:

1. Thermocouples
2. Resistance temperature detectors or RTDs
3. Thermistors
4. Diode temperature sensors or silicon bandgap temperature sensors
5. Liquid thermometers
6. Bimetallic thermometers
7. Infrared thermometers

The principles and operation of some of these sensors are described in the subsequent chapters. Among all the sensors, thermocouples and RTDs are the most

C.M. Jha (✉)
Intel Corporation, Santa Clara, USA
e-mail: cmjha75@gmail.com; Chandra.mohan.m.jha@intel.com

© Springer Science+Business Media New York 2015
C.M. Jha (ed.), *Thermal Sensors*, DOI 10.1007/978-1-4939-2581-0_1

Table 1.1 A comparison between thermocouple and RTDs

	Thermocouple	RTD
Temp. range	Wider	Narrower
Accuracy	Lower	Higher
Cost	Lower	Higher

Fig. 1.1 Transistor Node scaling with technology generation

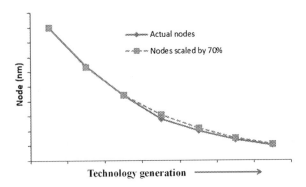

commonly used in industries and research laboratories. Thermocouples have large temperature ranges of less than −200 °C to more than 2000 °C and fairly good accuracy. RTDs can provide better accuracy compared to thermocouples, but they have a smaller temperature range and can be costlier than thermocouples. A comparison between thermocouple and RTDs is summarized in the Table 1.1.

For measuring junction temperatures in a silicon device, RTDs and diode temperature sensors are most commonly used. The foot-print, cost and accuracy of the sensor are taken into account while deciding the type of the sensor to be used in a silicon device. The effort is to minimize the die space occupied by the temperature sensors. In a microprocessor, the silicon die space is very valuable and designers would prefer to utilize most of the space for computing purposes while minimizing space for the temperature sensors. In addition, Moore's Law states that with every generation, the die size of the microprocessor shrinks (Moore 1965). Figure 1.1 shows the historical transistor node scaling with every technology generation. Transistor dimensions have been scaling down by a factor of 0.7x, which results in the doubling of the transistor density every generation. The scaling adds an additional challenge in the cost effectiveness of integrating temperature sensors on the die.

On the other hand, an error in the die temperature measurement can result in either over-design of the product, leaving the performance margin unutilized or under-design of the product, making it susceptible to thermal failures during its operation. The source of the error can be (1) accuracy, noise and precision of the temperature sensor and (2) the measurement methodology. Different types of temperature sensors exhibit different noise characteristics. There is a trade-off between accurate less noisy sensors and low cost sensors. Achieving both may be difficult and challenging. Miniaturization of the temperature sensors and their compatibility with the silicon fabrication process can sometimes provide benefits of both—more accurate sensor and lower cost. However, in many cases the requirements of the temperature sensors are mainly governed by their end applications. For example, a high performance computing device used in servers and workstations is limited by its silicon junction temperature whereas handheld devices like smartphones and tablets are generally limited by the outer skin temperature. The types of temperature sensors used in these scenarios can be distinct with different sets of accuracy and precision requirements and cost considerations. Similarly, in a research laboratory, the objective of the temperature measurement may be the accuracy of the sensors rather than their cost, whereas in the industry, the cost associated with the sensors becomes a primary factor in the decision making. Another important parameter is the response time of the sensor. For certain applications such as temperature control, the response time may be the driving factor in the sensor selection.

For a system designer or a thermal engineer, an overall understanding of thermal sensing and its application is necessary to make the right engineering decision. This book covers the topics that can provide the overall understanding of thermal sensing and applications. A brief description of the contents of some of the chapters is given below.

In Chap. 2, the reader will learn the fundamental principles of heat transfer, thermal sensors and their applications. The chapter describes heat transfer in a typical microprocessor package, different materials used in sensors, principles of sensor operation associated with accuracy, response time, noise, and reliability and the thermal sensors used in multiple segments including hand held segments. The reader will understand the fundamental aspects of the thermal sensors used in a typical semiconductor industry environment.

Chapter 3 explains the measurement capability of sensors in terms of the accuracy, repeatability and reproducibility of the sensors and the associated metrologies.

Chapter 4 presents the reasons why temperature measurement inside the microprocessor is important, how the temperature is currently measured, how the sensors are characterized, and how the junction temperatures are estimated. Readers will understand the importance of microprocessor junction temperature measurement and the advantages and challenges associated with it. Chapter 5 describes the future trend in the micro-electronics industry and the challenges associated with junction temperature sensing. Finally, Chap. 6 presents the specific application of temperature sensors in micro-energy converters.

Reference

Moore, G., "Cramming more components onto integrated circuits," Electronics, vol. 38, pp. 114–117, Apr. 19 (1965).

Chapter 2
Fundamentals of Thermal Sensors

Thu Huynh

Thermal sensors are found in many items, from commonplace items inside any home to more sophisticated applications. You can find sensors in household electronics like thermostats or thermometers. You will also find sensors in things as sophisticated as your personal computer or in a microprocessor. It is vital for processors to stay within the temperature range specification to perform reliably and for the processor to run at its expected speed performance.

In this chapter, we review the fundamental principles of heat transfer and describe heat transfer in a typical microprocessor package. We also touch on the principles of thermal sensors, including the various sensor materials, operation and applications in a typical semiconductor industry environment.

2.1 What Is a Thermal Sensor?

Sensors are devices that measure a physical or chemical reaction, such as volume flow or heat flux, through changes in electric resistance or signal (Kenny 2004). There are many types of sensors—flow, force, pressure, humidity and motion sensors are just a few. We are focused on one type of sensor in this book: thermal sensors.

2.1.1 Overview of Thermal Sensors

Temperature is the measure of the average kinetic energy of the molecules of a gas, liquid, or solid. A thermal sensor is a device that is specifically used to measure

T. Huynh (✉)
Intel Corporation, Santa Clara, USA
e-mail: thu.huynh@intel.com

© Springer Science+Business Media New York 2015

C.M. Jha (ed.), *Thermal Sensors*, DOI 10.1007/978-1-4939-2581-0_2

temperature. In this way, thermal sensors are able to give us a quantifiable way to describe the substance, whether it is an object, the environment in which an object is placed or the environment in which an object is distributed. More about how these sensors are applied to microprocessors are discussed in later chapters.

2.1.2 Types of Thermal Sensors

One well-known thermal sensor is a mercury or alcohol thermometer. It uses the volume of mercury or dyed ethanol, which expands when temperature increases, to measure temperature in a tube with a temperature scale. Though very well known, mercury and alcohol thermometers are not well suited to measure temperature in a personal computing device or microprocessor because they tend to be too large for those applications. Other kinds of thermal sensors that can be suited for personal electronics and microprocessors include thermocouples, resistance thermometers, silicon sensors and radiation thermometers.

2.1.2.1 Thermocouples

Thermocouples are sensors composed of two different metals at their sensing end. A voltage is created when there is a temperature gradient between the hot sensor element and the cold reference junction. The change in voltage can be reported as a temperature through the Seebeck effect (Love 2007). The Seebeck effect says that the change in voltage is linearly proportional to the change in temperature and the two variables are related to each other through a coefficient that is determined by the materials used in the thermocouple (Janata 2009). Figure 2.1 depicts the construction of a thermocouple.

2.1.2.2 Resistance Thermometers

Resistance thermometers are also known as resistance temperature detectors, or RTDs. They are typically made of a single pure metal (Dames 2008). Each metal has a material property of electrical resistance that is a function of temperature. The most accurate resistance thermometers are ones that use metals that have a very linear relationship with temperature, such as platinum. By using the relationship curves between electrical resistance and temperature, when the resistance of the metal is measured, a temperature can be calculated (Dames 2008). Figure 2.2 depicts the construction of one type of resistance thermometer.

2.1.2.3 Thermistors

A thermistor is a specific type of resistance thermometer. Thermistors are made of metal wires connected to a ceramic base made of several sintered, oxide semiconductors (Janata 2009). Like other resistance thermometers, the change in

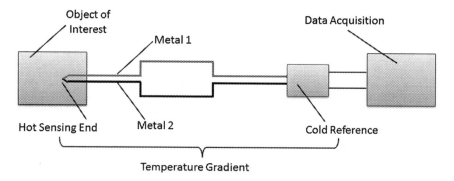

Fig. 2.1 Thermocouple construction. Adapted from Love (2007)

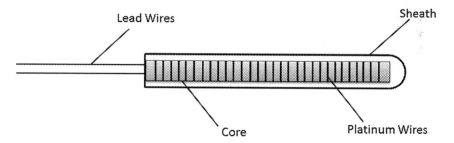

Fig. 2.2 An Example of resistance thermometer construction. Adapted from Desmarais and Breuer (2001)

temperature can be calculated from the change in resistance. But unlike traditional resistance thermometers, the relationship is not very linear. Thus, the temperature range in which thermistors can be used is small compared to traditional resistance thermometers. But thermistors have the advantages of being small in size, inexpensive to buy and very sensitive to temperature changes, so they can be ideal to use in many electronics applications (Janata 2009).

2.1.2.4 Silicon Sensors

These sensors are made of silicon, a semiconductor that is used as the base material for most electronic microprocessors. The process of manufacturing these electronic devices is a carefully controlled, high-volume manufacturing process that includes deposition, doping, and careful layering of metals, oxides, and insulators (Peterson 1983). By utilizing this manufacturing process, integrated circuit (IC) sensors can be created as their own sensor device (Desmarais and Breuer 2001). They can also be embedded inside microprocessors as diodes (Rotem et al. 2006). These types of sensors can have their own memory, can have direct output to meters and can convert signals to temperature readings without extra equipment (Desmarais and Breuer 2001).

2.1.2.5 Radiation Thermometers

All substances and objects emit thermal radiation when it is at a temperature higher than absolute zero (0 K or −273.15 °C). There is a relationship between temperature and radiation energy emitted that can be used to calculate the temperature of the object surface. Unlike other sensors discussed above, radiation thermometers are primarily used at a distance from the object of interest and can be used for hard-to-reach objects. An example of a radiation thermometer is an infrared camera, which measures infrared wavelengths that emit from an object.

2.2 Heat Transfer and Microprocessors

Power is an important design feature of microprocessors. It is linked to the expected silicon performance and also generates the heat that must be cooled from the part. The goal of microprocessor thermal management is to cool the processor efficiently within a specified temperature to ensure reliable performance over the lifetime of the part. Thus, it is important to understand the heat transfer mechanisms in the microprocessor that cool the processor in order to understand the importance of thermal sensors in microprocessors.

2.2.1 Fundamentals of Heat Transfer

Thermodynamics describes the fundamental behavior of heat and temperature and includes the three laws of thermodynamics. Heat transfer goes further and describes the mechanisms of heat exchange and the rate at which heat flows, giving us a way to calculate heat flow within, to and from objects or the environment. There are three modes of heat transfer: conduction, convection and radiation.

2.2.1.1 Conduction

Conduction is the first mode of heat transfer that we will discuss. It is a prominent mode of heat transfer in electronics cooling. Undergraduate transfer textbooks typically devote a large portion of the text to this topic and those books will be able to give a more comprehensive and in-depth discussion on this mode of heat transfer.

Definition of Conduction: Conduction is the heat transfer through solids. It can also occur with stagnant fluids. The one-dimensional rate of conductive heat transfer is determined by Eq. 2.1:

$$Q_{conduction} = \frac{kA(T_{hot} - T_{cold})}{L} \tag{2.1}$$

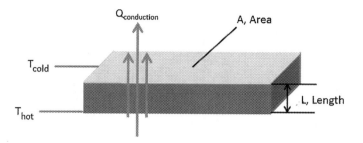

Fig. 2.3 Heat transfer through an object by conduction

where $Q_{conduction}$ is heat flow, k is the thermal conductivity of the material, A is the cross-sectional area of heat flow, T_{hot} is the temperature of the hot surface, T_{cold} is the temperature of the cold surface and L is the length of the material through which heat is conducting. Figure 2.3 depicts the heat transfer through a solid material by conduction. The different variables of the conduction heat transfer Eq. 2.1 are shown.

The conduction resistance is defined by Eq. 2.2:

$$R_{conduction} = \frac{L}{kA} \tag{2.2}$$

As shown in Eq. 2.2, in order to minimize the conduction resistance, conductivity of the material and cross-sectional area of material is maximized while the through-path (length) of the material is minimized.

Three-Dimensional Conduction: Conduction can also occur in three dimensions if the heat source size is smaller than the conducting material. If the heat source is smaller than the conducting material, the heat is concentrated in one spot, causing a hot spot. In Fig. 2.4, heat is being conducting in the x and y direction as well as the z direction. The conduction resistance in the x and y direction is also known as the spreading resistance, which can be solved in idealized boundary conditions by first order equations or more often through computational software. Figure 2.4 depicts one-dimensional and three-dimensional conduction through a solid, as well as the transition between the two types of conduction.

Contact Resistance: Another thing to consider in conduction is contact resistance between two different solid materials. When two materials join to form a conduction path, there is resistance at the point where the two materials join. This occurs because the two surfaces are rarely completely flat and voids of air are present at the junction of two surfaces. This must be taken into account when thermal solutions are applied to electrical packages or when an external thermal sensor like a thermocouple is attached to the package. Minimizing contact resistance is a key factor to consider in electronics packaging and their cooling.

Example 2.1 A rectangular block has an area of 400 mm². An engineer would like to use it to cool his heat source that is producing a total of 50 W with a specification of 90 °C. The heat source is placed in a chamber with an air temperature of

heat source sizes decreases, hot spots introduced, conduction becomes three-dimensional

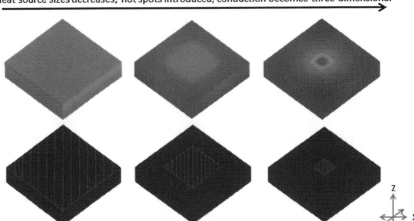

Fig. 2.4 Transitions between one-dimensional and three-dimensional conduction

65 °C. If the through-length is 10 mm, what is the minimum conductivity of the material in order to cool the heat source to specification using the block alone?

Solution

In this example, the mode of heat transfer is conduction through the block. The given information is annotated in the illustration below to help us solve this problem.

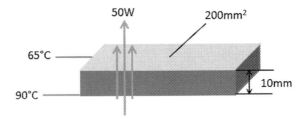

We will start with Eq. 2.1 and replace with the given variables. We want to solve for conductivity k in units of $W\, m^{-1}\, K^{-1}$.

$$Q_{conduction} = \frac{kA(T_{hot} - T_{cold})}{L}$$

In order to solve the equation with the correct units, the area and length units must be changed to meters, m. The known values are plugged into the conduction equation to give:

$$50\,W = \frac{k(0.0004\,m^2)(90\,°C - 65\,°C)}{0.01\,m}$$

Rearranging to solve for conductivity k:

$$k = \frac{(50\,\text{W})(0.01\,\text{m})}{(0.0004\,\text{m}^2)(90\,^\circ\text{C} - 65\,^\circ\text{C})}$$

$$k = 50\frac{\text{W}}{\text{mK}}$$

By re-ordering Eq. 2.1, we solve for k and find it to be $50\,\text{W}\,\text{m}^{-1}\,\text{K}^{-1}$.

2.2.1.2 Convection

Convection is the second mode of heat transfer we will discuss. Along with conduction, it is typically a large part of electronics cooling in active, fan-cooled systems. An undergraduate textbook can be consulted for in-depth information.

Definition of Convection: Convection is the heat transfer from a surface to a fluid. Some common fluids include air and water. Other fluids such as alcohol and oil can also be mentioned in cooling electronics. The rate of convection heat transfer is determined by Eq. 2.3 below:

$$Q_{convection} = h_c A \left(T_{ambient} - T_{surface} \right) \tag{2.3}$$

where $Q_{convection}$ is heat flow, h_c is the convection heat transfer coefficient, A is the surface area, $T_{ambient}$ is the ambient temperature of the fluid, and $T_{surface}$ is the temperature of the surface of the material. Figure 2.5 depicts the airflow profile and temperature profile you would expect through convection from the surface of the rectangular object. $T_{surface}$ is warmer than $T_{ambient}$, which corresponds to the lower airflow velocity at the surface.

The convection resistance is described by Eq. 2.4:

$$R_{convection} = \frac{1}{h_c A} \tag{2.4}$$

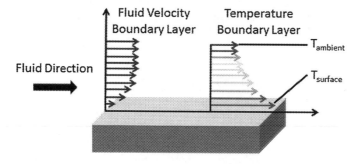

Fig. 2.5 Convection and its airflow and temperature boundary layers

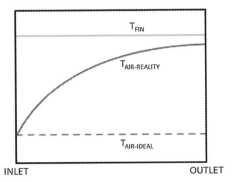

Fig. 2.6 Air heating: ideal versus real air temperature versus fin temperature

As shown in Eq. 2.4, to minimize convection resistance, the convection coefficient and surface area should be maximized.

Surface Area and Fins: To increase surface area and the effectiveness of convection heat transfer, fins are often added to the base of electronic cooling solutions. The fins are often made of a thermally conductive material such as aluminum or copper. Fins can be formed by extrusion from a large block or they can be formed and stacked on top of the base with solder.

Air Heating and Pressure Drop: Another consideration is the air heating along the length of the fin. In an idealized condition, the temperature difference between the air and the fin along the length of the fin would be constant. However, in reality, the temperature of the air increases along the length of the fin and the fin's ability to remove heat is diminished as the length of the fin is increased. It is not effective to simply make the fins infinitely long. Figure 2.6 shows the ideal versus real case for air heating. Additionally, longer fins will have a higher pressure drop across the fins compared to a heat sink with shorter fins. For heat sinks with longer fins, the total airflow will be reduced and thus cooling will be reduced for a given fan curve.

Optimizing Fin Performance: The fin stacks themselves have a fin resistance, which is a function of convection coefficient, fin efficiency and fin surface area. The balance among all three will determine the optimal fin geometry for a given set of boundary conditions, including cost considerations and manufacturing abilities.

2.2.1.3 Radiation

The last mode of heat transfer to discuss is radiation. Typically, this mode of heat transfer can be more complex than either conduction or convection. Graduate level classes cover this mode in more detail. In cooling microprocessors, radiation can play a large part when the primary mode of cooling is natural convection, with no active fan.

Definition of Radiation: Radiation is the heat transfer between surfaces via electromagnetic waves. All matter at a nonzero temperature emits electromagnetic waves, including gases and liquids. However, in many undergraduate heat transfer textbooks, the focus is only on solids. The rate of heat transfer is described by Eq. 2.5:

$$Q_{radiation} = h_r A (T_{source} - T_{surrounding}) \tag{2.5}$$

where $Q_{radiation}$ is the heat flow, h_r is the radiation heat transfer coefficient, T_{source} is the temperature of the source and $T_{surrounding}$ is the temperature of the surroundings (Bergman et al. 2011). The radiation heat transfer coefficient can also be described by Eq. 2.6:

$$h_r = \varepsilon \sigma (T_{source} - T_{surrounding})(T_{source}^2 + T_{surrounding}^2) \tag{2.6}$$

where ε is the emissivity of the source and σ (5.67×10^{-8} W m^{-2} K^{-4}) is the Stefan-Boltzmann constant (Bergman et al. 2011). The radiation heat transfer coefficient is similar to the convection heat transfer coefficient, but the radiation heat transfer coefficient is much more dependent on temperature as shown by raising the temperature terms in Eq. 2.6 to the third power (Bergman et al. 2011). The net radiation heat transfer is well known by Eq. 2.7, where h_r in Eq. 2.5 is replaced by Eq. 2.6:

$$Q_{radiation} = \varepsilon \sigma A (T_{source}^4 - T_{surrounding}^4) \tag{2.7}$$

Figure 2.7 depicts the heat transfer of an object through radiation. The variables in Eq. 2.7 are highlighted in the figure.

Components of Radiation Heat Transfer: An ideal radiative surface is called a blackbody with an emissivity of 1, but in reality, no surfaces are ideal. Radiative heat transfer is highly dependent on all the bodies surrounding the surface in

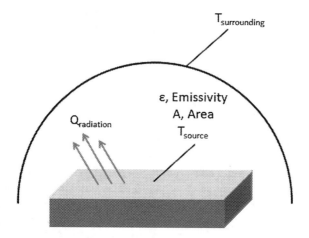

Fig. 2.7 Heat transfer through radiation from an object to its surroundings

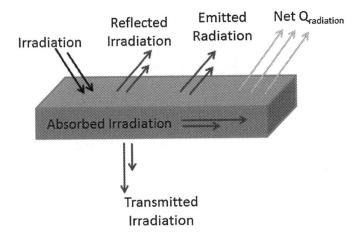

Fig. 2.8 Components of radiation heat transfer

question and the temperature and finish of the surfaces. A solid surface can emit, absorb, reflect, and transmit radiation depending on the material in question. For example, an opaque material can reflect radiation, whereas in a semitransparent material, radiation can be transmitted through the material. Unlike absorption and emission, reflection and transmission do not affect the total thermal energy of the material (Bergman et al. 2011). Figure 2.8 depicts the various components of heat transfer of an opaque solid through radiation.

2.2.2 Heat Transfer in a Microprocessor

As discussed in the previous section, the three modes of heat transfer can play important roles in the cooling and temperature sensing of microprocessors. Typically, in active, forced air-cooled systems, conduction and convection play the largest role. In natural convection systems or in systems where there is no room to attach a fan, radiation can play an important role.

2.2.2.1 Active Cooling

In active cooling systems, fans are attached in the system; they provide airflow over the microprocessors and can be used to help cool them. In these cases, the thermal engineer can conduct heat from the package out to a heat sink made of a thermally conductive material such as copper or aluminum. In cases where more cooling is needed, fins can be added to the cooling solution. Though actively cooled, it is also important to consider the conductive path from the package to the motherboard or any other substrate the microprocessor is attached to. Figure 2.9 depicts a few examples of heat sinks that can be used to cool microprocessors.

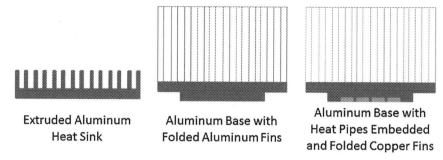

Fig. 2.9 Various heat sink configurations

2.2.2.2 Natural Convection Based-Cooling

In natural convection cooling solutions, no air movers are present or very little air is available due to airflow blockages. In these cases, radiation is an important cooling mechanism as well as conduction into the motherboard and natural convection. Analysis would have to include all the parts in the system in order to accurately predict thermal performance of the microprocessor or object in question. This analysis is one of the most complex heat transfer modes to model because all the surfaces, materials, and properties associated with the object must be specified accurately.

2.2.3 Package Thermals

Heat is spreading through conduction within the die and out of the package via active or natural convection cooling. The package is composed of the die, which is attached to a substrate made of FR-4 material with embedded copper layers. The package designer may decide not to cover the die; this is typically called a bare-die package. Alternatively, a package can include over-molding made of an epoxy or plastic to protect the die. It can also be covered with a copper integrated heat spreader (IHS) to help reduce spreading resistance and increase conduction to the heat sink. If the die is covered with a spreader, another material is needed between the die and spreader to fill in small air voids and gaps with conductive material. This interface material is typically called the thermal interface material. In Fig. 2.10, three package options (bare die, over-molded, integrated heat spreader) are shown.

Example 2.3 A heat sink is placed over a package with an integrated heat spreader. The package itself is attached to a substrate. The heat sink has metal fins and a metal base. Draw the resistance stack that should be accounted for in order to accurately find the die temperature (junction). Make sure to take into account any interfaces that would be present in a real assembly.

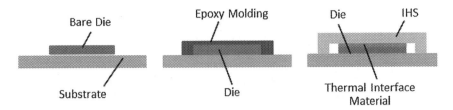

Fig. 2.10 Various package configurations with and without lid protection

Solution

To understand the different interfaces that should be accounted for, a schematic would be the first place to start. The problem statement gives the description for what the thermal solution looks like. The illustration below depicts the description given in this example of a heat sink with fins placed on top of a package with an integrated heat spreader.

The heat sink can be a radial heat sink or one of another shape, but for simplicity, we chose a simple rectangular base with straight fins. In this problem, it does not matter how tall or long the fins are. The base area and thickness are also not important. In this problem, the package is a ball-grid array (BGA) package, attached to the motherboard using solder balls. Alternatively, the package can be attached to the motherboard using a land-grid array (LGA) socket.

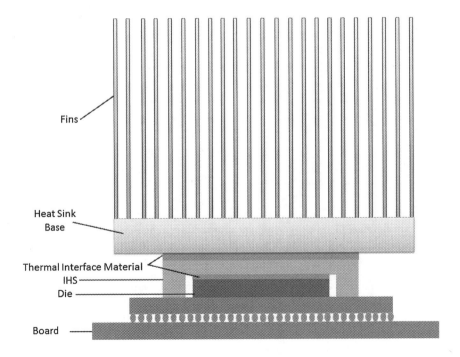

There are two places where thermal interface material would be placed in this configuration: within the package between the die and integrated heat spreader (Thermal Interface Material 1) and between the integrated heat spreader and the heat sink base (Thermal Interface Material 2).

Not only can power go up from the die into the heat sink, power may also go down through the substrate. It is also important to know that if power is dissipated into the substrate, the power will eventually be cooled by T_{air}.

The resistance stack can be created by choosing temperature points along the path through which heat will be dissipated, as shown in the next schematic.

The die temperature is also referred to as the junction temperature in many specification sheets and is shown as the junction temperature in this example.

It is important to note that the package resistance is in reality made of many parts: spreader resistance, thermal interface material 1 resistance, and so forth. However, many specification sheets simply give an overall resistance target from junction to case, which should be defined in the specification sheet, rather than describe all the details.

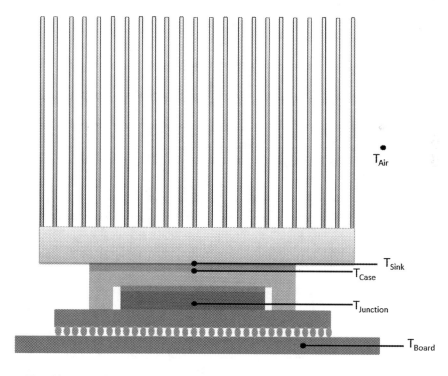

For this example, the resistance network is shown in the following illustration. This network reflects the simplification within the package by combining the spreader and thermal interface material 1 resistances into one resistance and reflects that the power through the package will be cooled by air.

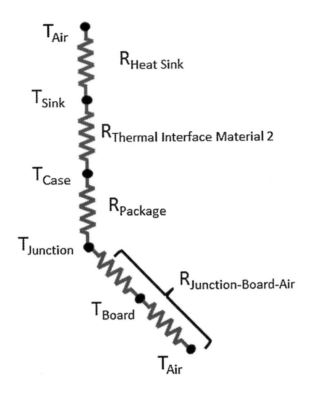

2.2.4 Need for Temperature Measurement in a Microprocessor

Each microprocessor needs a mechanism to give feedback to the user and system on how hot it is getting when it is powered. Without temperature measurement, the system may not provide enough airflow to cool the part or the engineer may not know if the thermal solution is adequate to keep the microprocessor within specification.

2.2.4.1 Hot Spots in the Microprocessor

In an ideal thermal situation, the microprocessor would be uniformly heated when it is powered. However, in reality, the microprocessor is made up of many different subsections that are designed to do different tasks. Thus, it is often the case that certain parts of the microprocessor are hotter than others and hot spots develop on the die where the microprocessor is more heavily utilized. Where the hot spots are located is dependent on the application and the layout of the die. The non-uniformity of heating and its dependence on workload application adds more complexity to understanding the processor temperature and it is not easily modeled accurately. Sensors are needed to understand where the hottest temperature is in real-use conditions.

2.2.4.2 Microprocessor Performance

There are many inputs into determining microprocessor performance, including voltage, leakage power, and thermal design power (TDP). The inputs are dependent on temperature and subsequently, the performance of the part, its frequency, is also dependent on temperature. Higher TDP, higher voltage, and lower leakage power increases frequency. This is the ideal case when performance is the most important variable in the design. However, if both temperature increases and reliability metrics are fixed, the voltage decreases and leakage power increases, leaving less power devoted to TDP. This is the opposite trend of what is needed for optimal performance. It is vital to measure the temperature of the microprocessor accurately to properly set frequency.

2.2.4.3 Microprocessor Reliability

The silicon microprocessor has an upper temperature limit specification that is set to prevent immediate microprocessor damage. In addition, microprocessors have an allowable failure rate and are specified to work for a certain period of time, usually in the number of years. If reliability is relaxed and the part can tolerate more failures or a shorter lifetime, voltage may not need to decrease when temperature increases, which allows for better performance. To ensure that there is minimal immediate damage and that the part failure rate is acceptable in its lifetime, the microprocessor temperature must be accurately understood.

2.3 Sensor Materials

There are many materials that are used for sensors. Metals are typically used for RTDs and thermocouples. The semiconductor silicon is used to make microprocessors and this is where many thermal sensors are built into the die. The following is a quick, but by no means an exhaustive, overview of the materials.

2.3.1 Platinum Sensors

Platinum is the choice material to use for RTDs. The typical temperature usage range for platinum RTDs is −250 to 600 °C, but some platinum RTDs can be used up to 850 °C (Love 2007). Its electrical resistivity has a very linear relationship

Table 2.1 Physical and thermal properties of platinum

Physical property	Platinum (Pt)
[a]Thermal conductivity (W m^{-1} K^{-1}) at 25 °C	71.6
[b]Melting point (K)	2042
[c]Temperature coefficient of resistance	3.92×10^{-3}
[d]Electrical resistivity Ω cm at 0 °C	9.6×10^{-6}
[b]Coefficient of linear thermal expansion (K^{-1})	8.8×10^{-6}
[e]Specific heat capacity (J g^{-1} K^{-1})	0.133
[b]Density (g cm^{-3})	21.45

Sources [a]Ho et al. (1972)
[b]Touloukian et al. (1975)
[c]Serway (1990)
[d]Hall (1968)
[e]Wagman et al. (1982)

with temperature over a large range of temperature (Dames 2008) and has a higher resistivity compared to other metals such as copper and nickel, which make it ideal for RTDs (The RTD 2014). The temperature coefficient of resistance (TCR) is 0.0039 K^{-1} at room temperature, where the TCR is the change in resistance per unit change in temperature. A higher TCR represents a more sensitive RTD to temperature and a high TCR is ideal. As a material, it is chemically inert and stable in different kinds of environments and is not likely to corrode or reduce, making it useful to measure temperature in many different environments (Dames 2008). Table 2.1 shows a few common thermal and electrical properties for platinum.

2.3.2 Thermocouple Materials

Many metals and alloys other than platinum are used in sensors, especially thermocouples. Though RTDs are mostly made of platinum, copper and nickel RTDs can also be found. In thermocouples, alloys are common materials (ANSI and IEC Color Codes 2014). Because of the wide range of materials, the temperature range of thermocouples is much wider than RTDs: temperatures can range as low as −270 °C and as high as 2300 °C. Common non-alloy thermocouple metals are summarized in Table 2.2.

In Table 2.3, a few common thermocouple alloys are summarized. The alloys in the table are made of copper, nickel, and chromium and they are found in J, K, T, and E thermocouple types.

Care must be taken in choosing thermo-electric materials for the environment they are used in: for example, thermocouples containing iron can be more susceptible to oxidization at high temperatures and is recommended for lower temperatures (ANSI and IEC Color Codes 2014). Chromium-based thermocouples can also see oxidation or "green rot" at higher temperatures when exposed to low levels of oxygen (Nicholas and White 2001). Table 2.4 summarizes the recommended and limiting conditions of a handful of thermocouple types.

Table 2.2 Physical and thermal properties of copper, nickel and iron

Physical property	Cu	Ni	Fe
[a]Thermal conductivity (W m^{-1} K^{-1}) at 25 °C	401	90.9	80.4
[b]Melting point (K)	1357.6	1728	1811
Temperature coefficient of resistance (Ω/Ω °C)	[c]3.9×10^{-3}	[d]5.9×10^{-3}	[c]5.0×10^{-3}
[e]Electrical resistivity Ω cm at 0 °C	1.545×10^{-6}	6.23×10^{-6}	8.7×10^{-6}
[b]Coefficient of linear thermal expansion (K^{-1})	16.5×10^{-6}	13.4×10^{-6}	11.8×10^{-6}
[f]Specific heat capacity (J g^{-1} K^{-1})	0.385	0.444	0.449
[b]Density (g cm^{-3})	8.933	8.90	7.87

Sources [a]Ho et al. (1972)
[b]Touloukian et al. (1975)
[c]Serway (1990)
[d]Electrical Conductivity (2014)
[e]Hall (1968)
[f]Wagman et al. (1982)

Table 2.3 Physical and thermal properties of copper-nickel and nickel-chromium alloys[a]

Physical property	Cu-Ni	Ni-Cr
Thermal conductivity (W m^{-1} K^{-1}) (typical)	20	19.2
Melting point (°C)	1260	1427
Temperature coefficient of resistance (Ω/Ω °C)	-0.01×10^{-3}	0.4×10^{-3}
Electrical resistivity Ω m (typical)	49×10^{-8}	70.6×10^{-8}
Coefficient of linear thermal expansion (K^{-1})	18.8×10^{-6}	13.1×10^{-6}
Density (g cm^{-3})	8.9	8.73

Source [a]Physical Properties of Thermoelement Material (2014)

Table 2.4 Recommended environmental conditions and thermocouple type[a]

Thermocouple type	Recommended and limiting conditions
E	Recommended for inert and oxidizing environmental conditions. Not recommended for use in reducing or vacuum environmental conditions
J	Recommended for inert, reducing and vacuum environmental conditions. Not recommended for use in high temperature, oxidizing environmental conditions
K	Recommended for inert and oxidizing environmental conditions. Not recommended for use in a reduced or vacuum environmental conditions
R	Recommended for inert or oxidizing environmental conditions without metal sheath protection
T	Recommended for inert, moist, oxidizing and vacuum environmental conditions

Source [a]ANSI and IEC Color Codes (2014)

Table 2.5 Physical and thermal properties of silicon

Physical property	Silicon (Si)
[a]Thermal conductivity (W m^{-1} K^{-1}) at 25 °C	149
[b]Melting point (K)	1687
[c]Specific heat capacity (J g^{-1} K^{-1}) at 300 K	0.712
[b]Density (g cm^{-3})	2.42

Sources [a]Ho et al. (1972)
[b]Touloukian et al. (1975)
[c]Hall (1968)

2.3.3 Silicon Sensors

Silicon is a semiconductor, which at a very basic level is a material that has an electrical resistance between an insulator and conductor (Yacobi 2002). Silicon has a diamond lattice structure and its neutral valence electron configuration allows silicon to equally share its valence electrons with other elements (Yacobi 2002). Because of this, silicon can be doped with other elements near it on a periodic table, such as boron or phosphorous, to fill its lattice structure with electrons that carry electrical current, thereby increasing the electric properties of silicon by orders of magnitude (Berger 2013). Thermally, silicon's thermal conductivity increases as temperature increases, the opposite of what occurs with most metals (Yacobi 2002). Most importantly, electronics are often made of silicon for several key reasons:

- Silica can be purified and turned in single-crystal silicon in a very pure form by vapor deposition (Habashi 2013).
- It is an inexpensive material (Peterson 1983).
- Its melting point is sufficiently high enough for silicon to be stable during high volume manufacturing, specifically for high temperature oxidation, diffusion, and annealing (Yacobi 2002).
- Silicon electronics can be manufactured in batches very precisely (Peterson 1983).

However, one important disadvantage of silicon sensors is the narrow range of temperature use compared to thermocouples and resistance thermometers: silicon sensors are generally only good from −50 to 150 °C (Bakker 2002). Table 2.5 is a summary of some of the electrical and thermal properties of silicon.

2.4 Principles of Thermal Sensors

There are many considerations to take into account in choosing a sensor to use. Some questions that frequently come up during that process are:

- What is being measured?
- What cost is acceptable?
- What is the accuracy required for the measurement?

In this section, we will have a quick overview of some factors that play a part in selecting a sensor for use in the industry, including temperature range for measurement, how the measurement will be made and the accuracy required.

2.4.1 Objective of Measurement

There can be several reasons why temperature is being measured for a given process, test, or object. The temperature may have a big effect on the chemical process that is being monitored. For example, in silicon high volume manufacturing, the temperature has to be set precisely to control layer growth and depth. In other cases, the temperature can be the main output of the test. An example of this case is during the performance test of a thermal heat sink. Temperature can also be measured to ensure reliability over time. Silicon dies usually have temperature monitors to ensure they remain under the maximum temperature limit and reliably work over a set number of years. Whatever the situation, care must be taken to pick a sensor that can get the job done accurately within the required conditions without being too costly.

2.4.1.1 Temperature Range of Measurement

It has been briefly mentioned in the section on sensor materials that each type of sensor has a certain temperature range of use. For thermocouples, the general range of use is between −270 and 1300 °C, but it can be as high as 2300 °C. For resistance thermometers, the temperature range of use is between −250 and 600 °C. For silicon sensors, the useful temperature range is between −50 and 150 °C. Of the three main sensor types, silicon has the most restrictive range, while thermocouples can be applicable in a wide variety of temperature cases.

Within the thermocouple group, different thermocouple types are rated for different temperature ranges. For example, Type J thermocouples, made with iron and nickel-copper materials, are useful to as high as 1200 °C. However, Type T thermocouples, made with copper and nickel-copper materials, are only useful to 400 °C (ANSI and IEC Color Codes 2014).

Resistance thermometers also have different temperature ranges depending on the material used. Platinum is the preferred choice of material for RTDs and its usable temperature range is widest, roughly between −200 and 650 °C. Nickel and copper alloys have the smallest usable temperature ranges, roughly between 0 and 205 °C, while nickel and copper have narrower usable temperature ranges between platinum and nickel/copper alloys (Desmarais and Breuer 2001). Thermistors, which use semiconductor oxides, have an even narrower range of use, between −100 and 300 °C (Desmarais and Breuer 2001).

If the conditions in which the thermometers are used reach or exceed the extremes of the usable temperature ranges, uncertainty can increase and temperature can deviate from the specifications. Determining the temperature ranges of the environments in which the sensors are being used during testing, processing and over their lifetimes will prevent these inaccuracies in measurement.

2.4.1.2 Environmental Conditions for the Thermal Sensors

Each type of sensor can also have limitations in certain environments. Material type will play a large part in this consideration. For example, iron and nickel-iron alloys can oxidize above 535 °C and corrode (Desmarais and Breuer 2001). These materials should be avoided under these conditions. Platinum is a very inert material and can be used in high temperatures and moist environments without any problem (Dames 2008). Each sensor with limiting materials can also be protected with special sheaths and materials under extreme environments, but this may change the size and cost of the sensor. In addition, some sensors, like thermocouples, would withstand higher shock and vibration conditions than thermistors or other resistance thermometers (Desmarais and Breuer 2001). In general, all environmental conditions—temperature, humidity, exposure to moisture, exposure to shock and vibration—must all be taken into consideration in choosing a sensor.

2.4.1.3 Cost

Cost can be an important factor in choosing a sensor. Platinum is an expensive material and because of the resistance of the material, it typically will result in a long sensor element (The RTD 2014). Copper, nickel, and lead are very low cost materials that can make them more appealing despite their material limitations, such as a narrower use-temperature range. Because silicon sensors can be produced in batches, their cost can also be quite low compared to resistance thermometers, but their temperature range is much narrower in comparison. In addition, the cost of the sensor can increase as the accuracy required increases. To approximately estimate the cost of a thermometer, including the costs of the material, manufacturing, calibration and metrology equipment needed, the following calculation can be used (Nicholas and White 2001):

$$Cost = \frac{\text{USD } 100}{Accuracy\ Required\ in\ °C} \tag{2.8}$$

Using Eq. 2.8, a thermometer requiring an accuracy of 0.01 °C will be 500 times more expensive than a thermometer requiring an accuracy of 5 °C, with the former costing an estimated USD 10,000. Thus, it is very important to not only pick the correct material, but the correct level of accuracy required to limit the cost of the thermometer.

2.4.1.4 Accuracy

Accuracy is the ability of the thermometer to exactly hit a specified temperature. This can be considered a qualitative description. In contrast, uncertainty is the range of expected error between the actual and ideal temperatures (Kenny 2004). Thermometer A can have an uncertainty of ±0.55 °C and thermometer B can have an uncertainty of ±1.3 °C quantitatively, while thermometer A can be described qualitatively as being more accurate than thermometer B.

Resistance Thermometers and Accuracy: Resistance thermometers can be accurate to 0.15 °C. Most platinum RTDs follow the International Electrotechnical Commission (IEC) standard 60751, which has equivalent standards published by the Deutsches Institut für Normung (DIN) and American Society for Testing and Materials (ASTM) (Dames 2008). The IEC accuracy standard is divided between Class A and Class B accuracy, where Class A is stricter and covers a smaller temperature range between −200 and 650 °C versus Class B, which has a range up to 850 °C. The formulations for temperature uncertainty are shown in Eqs. 2.9 and 2.10 (Dames 2008):

$$\text{Class A: } Uncertainty = \pm(0.15 + 0.002|T|) \text{ °C} \tag{2.9}$$

$$\text{Class B: } Uncertainty = \pm(0.3 + 0.005|T|) \text{ °C} \tag{2.10}$$

Table 2.6 shows the expected uncertainties calculated over a range of temperatures using Eqs. 2.9 and 2.10. Class B RTDs are easily found and bought and they will be less expensive than Class A RTDs. When buying RTDs, the class and required uncertainties should be specified.

Thermocouples and Accuracy: Thermocouples are generally less accurate than platinum RTDs. Thermocouples are classified according to type (J, K, and so on), where each type is distinguished based on the different metal combinations used to make the thermocouple. Manufacturers may follow the ASTM E230 specification listed in Table 2.7 for several thermocouple types. Some suppliers also have thermocouples that follow tolerances based on the IEC standard, which may have different tolerances from the ASTM E230 standard.

Even though manufacturers sell thermocouples of various types, it is always best to check them for accuracy in the environment they will be used in. Thermocouples do not always arrive within specification. Also note that care must be taken to specify which standard, type and temperature range is needed when ordering from suppliers.

Silicon Sensors and Accuracy: Silicon sensor accuracy is dependent on the process tolerances in manufacturing the silicon sensor. Uncertainties are also introduced during the conversion from analog to digital temperature as well as using power supplies with their own uncertainties (Sharifi et al. 2008). Typically,

Table 2.6 IEC accuracy standard—uncertainties of Class A and Class B

Uncertainty at temperature (°C)	Class A (°C)	Class B (°C)
−200	±0.55	±1.3
0	±0.15	±0.3
200	±0.55	±1.3
400	±0.95	±2.3
600	±1.35	±3.3
800	N/A	±4.3

Table 2.7 ASTM E230 standard: standard limits[a]	Type	Temperature range (°C)	Tolerance, whichever is greater
	E	0–900	1.7 °C or 0.5 %
	J	0–750	2.2 °C or 0.75 %
	K	0–1250	2.2 °C or 0.75 %
	R, S	0–750	1.5 °C or 0.25 %
	T	0–350	1.0 °C or 0.75 %

Source [a]ANSI and IEC Color Codes (2014)

the accuracy of silicon sensors is between 0.5 and 3 °C (Bakker 2002), though higher sensor errors have occurred in experience.

2.4.1.5 Response Time

In addition to cost, accuracy, and useful temperature range, the rate for the measured object or substance to change temperature should also be taken in account. For example, if the rate of temperature change for the substance in question is very fast, then the thermometer choice should also have a quick response. If the thermometer response is too slow, then it may inaccurately report the temperature at a given time and may allow the temperature to exceed a specified limit.

To get an accurate temperature reading, the thermometer should be in equilibrium with the system being measured. After heat is added, it will have to move by conduction or convection from the source closer to the thermometer and then be conducted into the thermometer—process steps that will all take time. Equation 2.11 shows the time it will take the thermometer to respond with system temperature change as a time constant, when the mode of heat transfer is conduction (Nicholas and White 2001):

$$\tau = C / \left(\frac{kA}{L} \right) \tag{2.11}$$

where C is the heat capacity of the fluid or object in J K^{-1} and the denominator is the inverse of the conduction resistance in Eq. 2.2. The time constant can also be written in terms of convection rather than conduction by using the inverse of the convection resistance in Eq. 2.4. Equation 2.12 would result (Tomsen 1998):

$$\tau = C / (h_c A) \tag{2.12}$$

The heat capacity is a material property that describes the amount of heat required to change the temperature of the object by one degree. It is found in units of J K^{-1}, which shows the relationship between heat, power and temperature. A larger heat capacity means that it will take more heat to change the temperature of a substance. Thus, generally, the higher the heat capacity of the object, the longer it will take for the temperature to change. Error between the thermometer and system will decrease exponentially with the time constant (Nicholas and White 2001).

Heat capacity is also function of mass. Oftentimes, the heat capacity is described in terms of a constant mass and the resulting specific heat capacity would be in units of $J\,kg^{-1}\,K^{-1}$. If the thermometer is large, the thermometer will have a longer response time. The thermometer will require more energy to change by a degree. Bigger thermometers like platinum resistance thermometers can have problems if quick response times are needed. However, small sensors like thermocouples will have quicker response times (Desmarais and Breuer 2001).

2.4.1.6 Calibration

The sensor's accuracy can be increased from the production or supplier specifications through calibration before test. Calibration can be done in the laboratory before test, or it can be completed by the manufacturer, usually at a cost. There are two basic ways that sensors can be calibrated, using fixed points or through bath calibration.

Calibration Using Fixed Points: In this context, fixed points refer to temperature points that are associated with thermodynamic properties of a pure substance, such as the triple point or melting point of a substance (Nicholas and White 2001). These points consistently occur at the same temperature and are highly reproducible, making them one of the most accurate ways to calibrate a thermometer. The International Temperature Scale of 1990 (ITS-90) is the adopted standard temperature scale that defines temperature in relation to fixed points. On the ITS-90 scale, the fixed points range from -270 to $1084\,°C$ in the original scale (Preston-Thomas 1990), with secondary points added for temperatures up to $3414\,°C$ (Bedford et al. 1996). One of the most important points widely used is the triple point of water, which occurs at $0.01\,°C$.

Bath Calibration: In bath calibration, the thermometer in question is calibrated against a standard or reference thermometer. Both are placed in the same environment, such as a bake chamber, a bath, or vacuum chamber (Dames 2008). Like in calibration using fixed points, it is important for both the thermometer to be calibrated and reference thermometer to be correctly attached or immersed in the chamber or bath. In addition, the reference thermometer must also be very accurate. In this type of calibration, many temperature points can be tested quickly and calibration can be completed for a batch of thermometers for a range of test temperature (Nicholas and White 2001). This kind of calibration can be completed for resistance thermometers, thermocouples, or silicon sensors.

2.4.2 Direct Versus Indirect Measurement

When a measurement is needed, another consideration to take into account is where the sensor is located. If the object is very far away, attaching a sensor directly on or in the object may be impossible. However, not attaching a sensor on an object of interest can introduce measurement errors. We briefly discuss direct and indirect measurements below, with a few examples of each.

2.4.2.1 Direct Measurement

In direct measurement, the sensor has direct contact with the object or substance that is being measured. Some examples of direct measurement include:

- Attaching a thermocouple to the center of an integrated heat spreader of a microprocessor to measure the case temperature of the package.
- Using a sheathed, platinum resistance thermometer to measure the bath temperature of water.
- Incorporating an on-die silicon diode onto a microprocessor to measure the temperature of the active area of the silicon.
- Placing thermocouples upstream and downstream of test setup, in the flow stream of the test, in order to measure the inlet and outlet air temperature of the test.

Direct measurement is often the most accurate way to take a temperature of an object. Even in direct measurement, making sure the correct location and correct number of sensors used is still important. For example, if the bath of water is taking up the volume of an Olympic-size pool, using one resistance thermometer may not capture temperature variation of different points in the pool of water.

2.4.2.2 Indirect Measurement

In indirect measurement, contact is not made with the object or substance of measurement. Radiation thermometers make almost exclusively indirect measurements with no contact with the object. However, thermocouple and resistance thermometers can also be making an indirect measurement if they are not placed exactly at the location of interest. Some examples of indirection measurement include:

- Using a radiation thermometer to measure the temperature of an object at a far distance.
- Placing an on-die diode that is on the die but not directly at the hot spot of the processor during workloads.
- Using an infrared camera to measure the temperature contours of a powered-on processor.

2.4.3 Temperature Scales

Temperature scales that are linked to heat sensitivity were not always in existence and it has taken many years to standardize the scales (Biró 2011). The first thermometers, where scales were included next to a measuring tube, are associated with several people, including Galileo and Ferdinand II of Tuscany (Biró 2011). For most scientists and engineers today, the most well-known scales are the Celsius, Fahrenheit, and Kelvin scale. In this section, we discuss these relevant scales. We also briefly discuss the more recent International Temperature Scale of 1990.

2.4.3.1 Fahrenheit

The Fahrenheit scale was named after Daniel Fahrenheit, who created this scale in 1724 (Biró 2011). This scale, like other scales of the time, is based on using fixed points to determine extremes of the scales and divided by an easy to remember number of steps (Nicholas and White 2001). Fahrenheit's scale is sometimes called the 96-based system, because he divided his scale by 96 parts. At the high extreme, he used the human body temperature of 96 °F as one of his fixed points and at the low extreme, he used the melting point of salty ice to be 0 °F (Biró 2011). This scale is often related to the Celsius scale, by the Eq. 2.13 below:

$$°F = 32 + \left[\frac{9}{5}\right] °C \qquad (2.13)$$

This equation is not always convenient. But most remember the relation by knowing that 32 °F is associated with the freezing point of water at 0 °C and that 212 °F is associated with the boiling point of water at 100 °C.

2.4.3.2 Celsius

The Celsius scale is also a well-known and widely used temperature scale. It is named after Anders Celsius, who created the scale in 1742 (Biró 2011). Like the Fahrenheit scale, it is based on using fixed points at the ends of the scale but instead of dividing by 96 parts, Celsius divided by 100 parts. For fixed points, he used the boiling temperature of water at the high end of the scale and the freezing temperature of water at the other end of his scale. Equation 2.11 can be rewritten to obtain degrees Celsius from degrees Fahrenheit by Eq. 2.14:

$$°C = \left[\frac{5}{9}\right] \times (°F - 32) \qquad (2.14)$$

2.4.3.3 Kelvin

The Kelvin scale was created to describe the absolute zero temperature point. Absolute zero is described as the point where there is no thermodynamic motion and where the thermal energy is zero (Biró 2011). Through experiments, absolute zero was determined to be −273.15 °C. Like the Celsius and Fahrenheit scales, the Kelvin scale is fixed at two points: 0 K is set to absolute zero and 273.15 K is set to the triple point of water. Because of how absolute zero and 0 K are defined, degrees Celsius and kelvin are often related by Eq. 2.15:

$$K = {}^{\circ}C + 273.15 \qquad (2.15)$$

Unlike degrees Celsius and degrees Fahrenheit, kelvin is not reported as degrees Kelvin. Instead, it is just report as K, a unit of temperature (Nicholas and White 2001).

2.4.3.4 ITS Scale

As mentioned earlier under calibration with fixed points, the International Temperature Scale of 1990 (ITS-90) is an adopted standard temperature scale that defines temperature in relation to fixed points. On the ITS-90 scale, the fixed points are from −270 to 1084 °C in the original scale (Preston-Thomas 1990). An addition with quality secondary points was added for temperatures up to 3414 °C (Bedford et al. 1996). The fixed points, which occur consistently at the same temperature and are highly reproducible, are chosen from thermodynamic points— boiling point, melting point, or triple point, to name a few—of substances such as water, hydrogen, copper, and more. On the original scale, the lowest fixed point was the vapor pressure point of helium at −270.15 to −268.15 °C and the highest point was the melting point of copper at 1084.62 °C. With the secondary reference points, the lowest fixed point is now zinc at −272.3 °C and the highest point is the melting point of tungsten at 3414 °C. The ITS scale is often used to provide a reference calibration point.

2.5 Sensing Noise

Sensors are manufactured with specified accuracy ranges, which can further be calibrated by the manufacturer or user to gain even more accuracy. Despite the knowledge and experience on obtaining accurate sensors, readings can still have inaccuracies due to noise. Noise can be caused by process, metrology, or it can be intrinsic to the sensor. This section briefly discusses those three sources of noise.

2.5.1 Process Noise

Many manufacturing processes are now very controlled and products from the manufacturing line are consistent and are mass copies of one another. Despite this control, it is impossible to hit the same values, such as thickness or diameter, every single time for every single product off the line. Manufacturing processes can have some standard deviation from the mean or set point, which the end user can see as ranges of accuracy in sensors or as sensors that are not made to an exact specified length or diameter.

A good example of process noise occurs in silicon manufacturing. This variation will be present in diodes that are embedded in the microprocessor or in stand-alone silicon sensors. To make silicon sensors and microprocessors, the manufacturing process involves growing and laying down several layers of silicon, oxides, and metals, each with its own thickness set point. If the layer is supposed to be 10 microns thick, it likely can vary between 10, 10.1, or 9.9 μ, depending on the tolerances allowed.

Sensors can be rated to be accurate to ± 5 °C, but a specific sensor could be accurate to 1 °C or perhaps 4 °C. When sensors are delivered from the manufacturing plant, process noise will be present. Calibration or tests versus temperature references (that is, a well-stirred water bath) can be used to understand how process noise may affect the performance of the sensors and to reduce any errors.

2.5.2 Metrology Noise

In addition to process noise, metrology noise is introduced through calibration and measurement equipment, such as chambers, power supplies and data acquisition machines, to name a few sources.

During Calibration: In order to calibrate sensors, the user or manufacturer must use reference points by calibrating to fixed points or reference thermometers. If the reference thermometer is not exactly accurate itself or if it has drifted over time, then the sensor being calibrated to it can see this error in accuracy. When baths or chambers are being used, the temperature distribution within the bath and chamber can introduce errors to the sensors being calibrated in them. If the temperature at the edge of the chamber is 0.2 °C warmer or cooler than the center, sensors near the edge would not be calibrated to the same temperature as those located in the center and a 0.2 °C error is introduced. It is best to make sure the bath and chamber is well-stirred and that the temperature distribution is as even as possible during calibration. It is also advisable for the reference sensors to be calibrated on a periodic basis.

During Measurement: During measurement, power supplies, data acquisition machines, temperature readers, voltmeters, multi-meters and similar equipment can be used in providing power, in recording data and in reading out data. These electronics will have specific tolerance ranges and accuracy limits separate from those of the sensors. Calibration may also be required for the electronics used in gathering measurements from the sensors. If, for example, the power supply is deviating slightly from the specification and is providing more voltage than is being reported, the sensor can read a higher temperature than it would at the correct voltage. It is advisable to use calibrated electronics with the sensors and to compensate for any errors introduced if needed.

2.5.3 Sensor Intrinsic Noise: Material and Thermal Effects

In all thermal sensors, there is noise that is associated with how the sensors work, with the types and quantity of materials used, and with the long periods of heat exposure. This kind of noise can be unavoidable, but can be reduced to minimize its effects. Ignoring or being unaware of these noise factors may introduce large errors in measurement.

Self-Heating: Self-heating errors occur when the temperature reading in the sensing element is increased due to heat caused by the sensing element itself. For resistance thermometers, the current flowing through the wires causes self-heating. The increase in temperature is described by Eq. 2.16,

$$\Delta Temperature_{self-heat} = \frac{R \times I^2}{D} \tag{2.16}$$

where R is the resistance of the element, I is the current flowing through the element, and D is a constant that describes the dissipation between the element and the fluid or solid being measured (Nicholas and White 2001). The relationship between the self-heat error and current tells us that even a small current can produce a large error because current is raised to the second power. To reduce self-heating effects, the current should be as small as possible while balancing the required accuracy of the temperature reading (Dames 2008). Self-heating effects can also be decreased by maximizing the constant D as much as possible—in order to do this, the contact between the sensor element and fluid or solid should be very good, with no gaps and no weak attachment methods. Despite the efforts to decrease self-heating errors, in some cases it will be impossible to avoid them—for example, in cases where the sensor is measuring the temperature of very stagnant air (Nicholas and White 2001). Stagnant air would not provide very effective cooling and it is likely that the whole sensor setup would self-heat in this environment.

Thermoelectric Effects: Thermoelectric errors occur when more than one type of metal is used in the construction of the sensing element and associated body and wiring (Dames 2008). Other areas where different metal wires can be introduced include metrology electronics, such as data acquisition machines, and power supplies. When more than one metal is used, the different metals can produce a Seebeck voltage that would lead to a temperature gradient along the sensing element and associated wires (Dames 2008). For resistance thermometers, this would lead to different voltage readings along the wire and introduce error to the temperature reading of the sensing element. The Seebeck effect is vital in order for thermocouples to work properly, but introductions of different metal types in the construction of thermocouple extension wiring can produce temperature gradients in the wires in a similar manner to what occurs in resistance thermometers (Nicholas and White 2001). In both sensor types, the degree of error is dependent on the Seebeck coefficient of the metals involved. The temperature gradient in the wiring can be as much as 10 °C and consequently, this temperature gradient can lead to inaccurate readings of 1 °C or more for resistance thermometers and thermocouples (Dames 2008); Nicholas and White 2001).

Other Thermal Effects: The long-term exposure to heat has the effect of elastically deforming the wires or leads of the sensor. With deformation, the metal can stretch and contract through several cycles, introducing errors as a result of different rates of expansion and contraction and the subsequent change in the resistance of the wire (Nicholas and White 2001). In addition, other causes of thermal gradients than the two previously discussed (self-heating and thermoelectric effects) will produce noise. For example, thermal effects that lead to larger thermal gradients can be introduced from inadequate connection between sensing wires and leads or extension wires (Nicholas and White 2001).

2.5.4 Sensor Intrinsic Noise: Lead Wire Resistance

For thermocouples, if the correct instrumentation is used with little current through the circuit, there should be negligible resistance in the wires (Nicholas and White 2001). Lead wire resistances have a larger effect on resistance thermometers, because current needs to flow through the sensors for RTDS to work. Resistance thermometers have several wire configurations with different numbers of wires: two-wire, three-wire, or four-wire. The circuit they are based on is commonly called the Wheatstone bridge (Love 2007). One of the resistors in the circuit is the resistance thermometer of interest. The other resistors are defined to complete the circuit in equilibrium at a reference temperature (Love 2007).

Two-Wire Resistance Thermometers: A two-wire configuration is a common and inexpensive configuration for resistance thermometers. However, it is also the configuration that is most susceptible to error. Figure 2.11 depicts the configuration of a simple two-wire bridge.

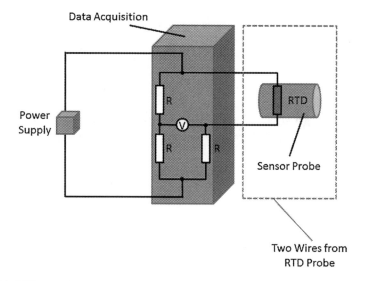

Fig. 2.11 RTD—simple two-wire resistance network. Adapted from Love (2007)

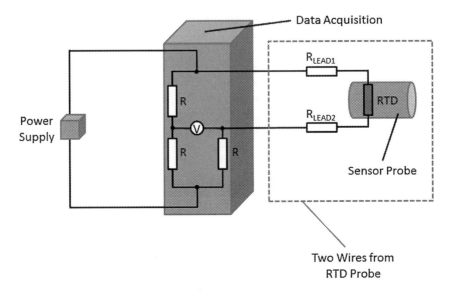

Fig. 2.12 RTD—two-wire resistance network with long leads. Adapted from Love (2007)

The bridge works well if the leads to the resistance thermometer are negligible and there is no measurable resistance in those leads. Error is introduced when the leads to the thermometer are very long, so that the lead resistances are not equal or small (Love 2007). Figure 2.12 depicts where the extra resistors would be introduced in series with the resistance thermometer. The extra resistances would cause inaccurate readings for the resistance thermometer.

Three-Wire Resistance Thermometers: A three-wire resistance thermometer features three leads from the resistance thermometer, as depicted in Fig. 2.13. It is more accurate but also more costly than a two-wire circuit. R_{LEAD1} and R_{LEAD3} will cancel out in this configuration when the circuit is in equilibrium (Love 2007). R_{LEAD2} is the resistance between the voltmeter and the lead across the circuit. The network should have very little current through R_{LEAD2} when the circuit is balanced so its resistance should be very small (Love 2007). In real conditions, there could be some error due to the lead resistances because no two resistors are exactly equal, but the error in a three-wire circuit is much less than using a two-wire circuit.

Four-Wire Resistance Thermometers: The most accurate circuit configuration for a resistance thermometer is the four-wire circuit, as depicted in Fig. 2.14. It is the most expensive because it uses more material than a two-wire or three-wire circuit. In this configuration, the lead resistances carry constant current in the outer loop and another inner loop with two other resistors measure the resistance across the thermometer directly (Love 2007).

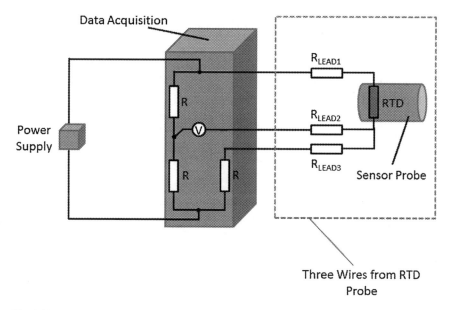

Fig. 2.13 RTD—three-wire resistance network. Adapted from Love (2007)

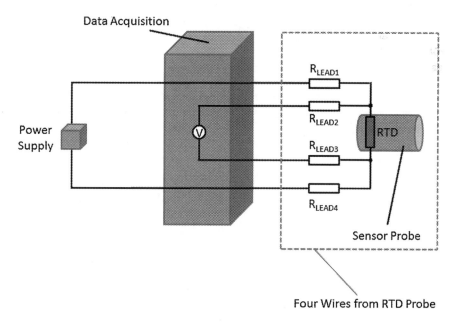

Fig. 2.14 RTD—four-wire resistance network. Adapted from Love (2007)

2.6 Sensor Reliability

Thermal sensors will be exposed to temperature cycling and high temperatures throughout their use. Inevitably, this will cause degradation in sensor accuracy and complete breakdown of the sensor as it reaches the end of its life. In addition, there are other sources that can cause complete sensor error or breakage well before the end of its expected lifetime. There are techniques and sensor configurations that can prevent catastrophic failure and extend its life, but sensors are not meant to last forever. This section discusses a few sources that contribute to unreliable accuracy and causes of sensor failure.

2.6.1 Shock and Vibration

Two causes of sensor failure and sensor reading drift are shock and vibration during packaging and shipping from the supplier to the calibration laboratory and ultimately to the use location. Sensors can also experience shock and vibration effects during attachment and during use while measuring the fluid or object temperature. Shock and vibration can cause total failure by breaking the sensor wires completely and cause slow deterioration and accuracy drift if the effects are experienced over time, for example, as a sensor that is attached to vibrating machinery would (Nicholas and White 2001). Shock and vibration can also cause the sensor wires to kink, bend or change shape, leading to changes in electrical resistance (Nicholas and White 2001). In addition, any metal strengthening due to plastic deformation caused by shock and vibration can change the electrical resistance of the wire (Nicholas and White 2001). To prevent shock and vibration effects, the sensors should be adequately packaged and carefully handled during shipping. Insulation can also be used to dampen the shock and vibration effects during shipping as well as use.

2.6.2 Hysteresis

Hysteresis is a condition where, in addition to the current sensor environment, the previous sensor environment also affects the current sensor readings (Kenny 2004). Hysteresis becomes apparent for temperature sensors during temperature cycling (Nicholas and White 2001). The metal wires used in sensors will stretch and contract as the temperature cycles between temperatures. The different rates of expansion and contraction will affect the next sensor reading and result in erroneous readings. As an example, let us take a sensor that is used to measure a fluid at a high temperature of 400 °C and is then used to measure fluid temperature as its temperature decreases. The sensor at 375 °C will likely not have had enough time to fully contract and will still see expansion effects from being exposed to 400 °C. The reading at 375 °C will subsequently have errors. As the number of cycles increase, the error also increases. Hysteresis effects can be reduced by allowing the sensor to contract back and reach equilibrium at a specific measurement point (Nicholas and White 2001). However, after many thermal expansions and contractions, the sensor will reach a point where deformation will no longer be elastic and the sensor will deform permanently. At that point, the sensor will be irreversibly damaged and its resistance will be permanently changed.

2.6.3 Chemical Environment

Because many thermal sensors and sensor peripherals are made of metals or alloys, the chemical environment to which the sensors are exposed becomes a very important factor for long-term reliability. Several different of types of thermocouples were described in an earlier section of this chapter, including ones made of iron, nickel, platinum, nickel-iron, copper-nickel, and other alloys. Each metal has a different reaction to moisture, vacuums, oxidizing, and reducing environments. For example, iron-based thermocouples may not be suitable at high-temperature, oxidizing environments because iron is easily oxidized (ANSI and IEC Color Codes 2014). In addition to the wires themselves, thermocouple wire insulators are subject to their own behaviors in different chemical environments. For example, many materials used as insulators for thermocouples break down in reducing environments, leaving the bare wires exposed to high temperatures (Nicholas and White 2001). Chemical environment effects lead to electrical resistive changes and to changes to the Seebeck coefficient of the metal materials (Nicholas and White 2001). This results in lower accuracies and can ultimately break the sensor. To avoid negative environment effects, thermal sensor type must be carefully chosen.

2.6.4 General Thermal Effects

Because thermal sensors are used to measure temperature, sometimes at very high temperature, thermal sensors are naturally going to suffer from thermal effects. The thermal effects are often compounded because of sensor exposure to unfit chemical environments or to vibration at the same time, as was already described earlier in this section. High temperatures will cause the metals to generally change shape, internal electron structure, and even external physical dimensions, which all change the sensor's resistance properties (Nicholas and White 2001). As temperature increases, the metals within one sensor may have different thermal expansion rates, causing work hardening (Dames 2008). High temperatures can also lead to metal migration in the wires, leading to metal contamination and increasing electrical resistance (Nicholas and White 2001). Because of these effects, it is important for the sensors to have proper insulation and for them to be used in the proper temperature environments. If sensors are not chosen carefully for measuring capability, the sensors may fail or result in erroneous measurements. Additionally, sensors should go through reliability tests to ensure that the thermal effects are understood before permanent installation.

2.7 Thermal Sensors in Handheld Devices

Smartphones and tablets have become very popular devices and they are present in many households in replacement of traditional laptops or landline phones. Because of widespread smartphone and tablet use, there has been a push to use the handheld devices to do more than their traditional functions of making phone calls, text messaging, or browsing the Internet. The ability of smartphones and tablets to connect to the Internet or to a wireless communication network also makes them good candidates for use as devices that collect information via sensors and share them with other applications or the Internet for consumption (Fujinami et al. 2013).

In handheld devices today, there are many embedded sensors that are commonly used. Accelerometers are used to detect which direction the handheld device is being held and they are also used in connection with applications that track running speed. There are also sensors that detect sound, which can be used to detect surrounding noise like music for music-recognition applications and light, which is frequently used today to automatically adjust screen brightness to accommodate changing dark or bright ambient lighting. In addition, there are GPS, proximity sensors, gyroscopes, and compasses embedded into smartphones that enhance location-based applications (Lane et al. 2010).

Within the handheld processor and attached to the board of the phone, thermal sensors can be used in traditional roles of monitoring the temperature of the processor and electrical hardware. These sensors are typically built-in silicon diodes or thermistors. However, despite the multi-functionality of smartphones and tablets, there are not as many instances where embedded sensors are being commercially used to monitor environmental readings such as temperature, humidity, or UV-radiation (Fujinami et al. 2013). Studies and development opportunities exist to expand the use of handheld devices to encompass thermal sensors. For example, there is a published proposal by Fujinami et al. to use smartphones as devices that can monitor temperature for heat stroke prevention by using a silicon sensor and Android†-based software (Fujinami et al. 2013). However, because handheld devices can be placed in several different positions, the inaccuracy of the temperature readings to the intended measurements can be large. For example, a handheld device placed in a bag will be more inaccurate when compared with a device hanging from a person's neck in measuring the ambient environment temperature (Fujinami et al. 2013). The complication of inconsistent handheld device placement would also make it difficult to use handheld devices to monitor body or skin temperature for medicinal purposes. Some other considerations for smartphone and tablet temperature monitoring to be effective include development of a robust and standard sensing methodology with handheld devices and agreement on where data is being stored and how it is transmitted (Lane et al. 2010). In addition, if thermal sensors are to be used to monitor health and ambient conditions uninterrupted, CPU bandwidth, memory use, and battery consumption with other applications in the handheld device should be considered and will need be balanced (Lane et al. 2010).

2.8 Thermal Sensors in Remote Applications

As sensor technology becomes more sophisticated, smaller, and cheaper, there have been research and proposals to use sensors wirelessly to measure an object or fluid from afar or to embed them into textiles to provide on-body sensing. In this section, we briefly discuss thermal sensors as wireless sensors and in wearables. Like the previous section on thermal sensors in handheld devices, the discussion is not exhaustive but highlights the emergence and development of how thermal sensors can be used in the future.

2.8.1 Smart Sensors

At the basic level, wireless sensors and sensors for wearables can be considered smart sensors. Traditional sensors are hard-wired, specialized to one task, localized, and require support equipment to transform the electrical input to the intended output like temperature (Mekid et al. 2010). Smart sensors can have the ability to collect various types of sensor data in one unit or die, like temperature, pressure, and humidity (Roozeboom et al. 2013). Smart sensors can also integrate various other functions to one sensor unit, including the ability to transform the input to a consumable output within the wireless unit, compensate for expected errors, and communicate and transmit data (Mekid et al. 2010). For temperature sensors, they can be made "smart" by integrating the analog-to-digital converter (ADC) with the temperature sensor (Bakker 2002). Temperature sensors are also likely to be silicon-based, like doped-Si resistance thermometers or bipolar transistor-based thermal sensors (Roozeboom et al. 2013).

2.8.2 Wireless Sensors

The temperature sensor can be one node in a network of many temperature sensing or other sensing nodes (Mukhopadhyay 2013). For wireless sensors, communication is typically transmitted through radio frequency signals. Currently, three of the more popular signals to use are Wi-Fi†, Bluetooth†, and Zigbee† (Mekid et al. 2010). The radio frequency should be chosen for the application so that there is minimal interference from other radio sources, like microwaves or television, and so that the wireless sensor can operate no matter the location (Mukhopadhyay 2013). Additionally, the cost and power-bandwidth of each type of signal should be considered. For example, Wi-Fi has large range and more data bandwidth capability, but it also consumes large amounts of power compared to using Zigbee signals (Mekid et al. 2010). There is an increasing interest in sensors to wirelessly monitor for applications such as environmental efficiency, transportation, in smart homes and in health care (Mukhopadhyay 2013). It is easy to envision the latter two areas as examples where temperature sensors can easily be applicable. Temperature sensors may be used to coordinate room temperatures within the home for increased energy efficiency or comfort. Temperature sensors can also be useful for health-care monitoring to measure body and skin temperature.

2.8.3 Smart Sensors for Wearables

For health-care monitoring, smart sensors can be embedded in a wearable, which can be defined as anything from traditional clothing or small electronics like wristwatches, wristbands, glasses or chest-bands (Anliker et al. 2004). There have been several attempts to use temperature sensors to monitor body temperature, such as in the advanced care and alert portable telemedical monitor (AMON) project (Anliker et al. 2004), in the Bioharness† monitoring system (Johnstone et al. 2012), and in protective equipment to protect firefighters (Talavera et al. 2012). Even though there is high interest in monitoring temperature for health care, the AMON and Bioharness monitoring systems have shown inaccuracies in using wearable temperature sensors. For the AMON system, the wristwatch approach does not provide an adequate means of predicting body temperature (Anliker et al. 2004). For the Bioharness chest-belt, readings from an infrared camera have low correlation to actual body temperature measurement by a calibrated thermistor attached to the skin (Johnstone et al. 2012). While these studies are not exhaustive, it does showcase some attempts at using temperature sensors in wearables and is also indicative of needed improvements in the future to provide accurate temperature readings for wearables.

References

Anliker, U., Ward, J. a, Lukowicz, P., Tröster, G., Dolveck, F., Baer, M., Keita, F., Schenker, E.B., Catarsi, F., Coluccini, L., Belardinelli, A., Shklarski, D., Alon, M., Hirt, E., Schmid, R., Vuskovic, M.: AMON: A Wearable Multiparameter Medical Monitoring and Alert System. IEEE Trans. Inf. Technol. Biomed. 8(4), 415–27 (2004).
Bakker, A. (2002): CMOS smart temperature sensors—An Overview. Proceedings of IEEE Sensors 2002. pp. 1423–1427. IEEE (2002).

Bedford, R.E., Bonnier, G., Maas, H., Pavese, F.: Recommended values of temperature on the International Temperature Scale of 1990 for a selected set of secondary reference points. Metrologia. 33, 133–154 (1996).

Berger, L.I.: Properties of Semiconductors. In: Haynes, W.M. (ed.) CRC Handbook of Chemistry and Physics, 94th Edition. pp. 12–80 – 12–93. CRC Press (2013).

Bergman, T.L., Lavin, A.S., Incropera, F.P., Dewitt, D.P.: Fundamentals of Heat and Mass Transfer, Seventh Edition. John Wiley & Sons (2011).

Biró, T.S.: How to Measure Temperature. In: Biró, T.S. (ed.) Is There a Temperature? Conceptual Challenges at High Energy, Acceleration and Complexity. Fundamentals Theories of Physics, vol 171,. pp. 5–27. Springer New York, New York, NY (2011).

Dames, C.: Resistance Temperature Detectors. In: Li, D. (ed.) Encyclopedia of Microfluidics and Nanofludics. pp. 1782–1790. Springer US (2008).

Desmarais, R., Breuer, J.: How to Select and Use the Right Temperature Sensor. Sensors Online, http://archives.sensorsmag.com/articles/0101/24/index.htm, (2001).

Fujinami, K., Xue, Y., Murata, S., Hosokawa, S.: A Human-Probe System That Considers On-body Position of a Mobile Phone. In: Streitz, N. and Stephanidis, C. (eds.) Distributed, Ambient, and Pervasive Interactions - First International Conference, DAPI /HCII 2013. Lecture Notes in Computer Science, vol. 8028. pp. 99–108. Springer Berlin Heidelberg (2013).

Habashi, F.: Silicon, Physical and Chemical Properties. In: Kretsinger, R.H., Uversky, V.N., and Permyakov, E.. (eds.) Encyclopedia of Metalloproteins. pp. 1998–2000. Springer New York (2013).

Hall, L.A.: Survey of Electrical Resistivity Measurements on 16 Pure Metals in the Temperature Range 0 to 273K. NBS Tech. Note 365. February, 1–111 (1968).

Ho, C.Y., Powell, R.W., Liley, P.E.: Thermal Conductivity of the Elements. J. Phys. Chem. Ref. Data. 1(2), 279–421 (1972).

Janata, J.: Thermal Sensors. Principles of Chemical Sensors. pp. 51–62. Springer US (2009).

Johnstone, J. a, Ford, P. a, Hughes, G., Watson, T., Garrett, A.T.: Bioharness(TM) Multivariable Monitoring Device. Part. I: Validity. J. Sports Sci. Med. 11(3), 400–8 (2012).

Kenny, T.: Sensor Fundamentals. In: Wilson, J.S. (ed.) Sensor Technology Handbook. pp. 1–20. Elsevier Science & Technology (2004).

Lane, N.D., Miluzzo, E., Lu, H., Peebles, D., Choudhury, T.: A Survey of Mobile Phone Sensing. IEEE Commun. Mag. 48(9), 140–150 (2010).

Love, J.: Temperature Measurement. In: Love, J. (ed.) Process Automation Handbook - A Guide to Theory and Practice. pp. 99–106. Springer US (2007).

Mekid, S., Starr, A., Pietruszkiewicz, R.: Intelligent Wireless Sensors. In: Holmberg, K., Adgar, A., Arnaiz, A., Jantunen, E., Mascolo, J., and Mekid, S. (eds.) E-maintenance. pp. 83–123. Springer London (2010).

Mukhopadhyay, S.C.: Wireless Sensors and Sensors Network. In: Mukhopadhyay, S.C. (ed.) Intelligent Sensing, Instrumentation and Measurement. Smart Sensor, Measurement, and Instrumentation, vol. 5. pp. 55–69. Springer Berlin Heidelberg (2013).

Nicholas, J. V., White, D.R.: Traceable Temperatures - An Introduction to Temperature Measurement and Calibration, Second Edition. John Wiley & Sons (2001).

Peterson, K.E.: Silicon as a Mechanical Material. Microelectron. Reliab. 23, 403 (1983).

Preston-Thomas, H.: The International Temperature Scale of 1990 (ITS-90). Metrologia. 27(3), 3–10 (1990).

Roozeboom, C.L., Hopcroft, M.A., Smith, W.S., Sim, J.Y., Member, S., Wickeraad, D.A., Hartwell, P.G., Pruitt, B.L.: Integrated Multifunctional Environmental Sensors. J. Microelectromechanical Syst. 22(3), 779–793 (2013).

Rotem, E., Hermerding, J., Aviad, C., Harel, C.: Temperature Measurement in the Intel Core Duo Processor. Procceedings of the 12th International Workshop on Thermal Investigations, Therminic, Nice 2006. EDA Publishing Association (2006).

Serway, R.A.: Physics: For Scientists and Engineers with Modern Physics, 3rd Edition. Saunders College Publishing, Philedelphia (1990).

Sharifi, S., Liu, C., Rosing, T.S.: Accurate Temperature Estimation for Efficient Thermal Management. 9th Int. Symp. Qual. Electron. Des. 137–142 (2008).

Talavera, G., Martin, R., Rodríguez-alsina, A., Garcia, J., Fernández, F., Carrabina, J.: Protecting Firefighters with Wearable Devices. In: Bravo, J., López-de-Ipiña, D., and Moya, F. (eds.) Ubiquitous Computing and Ambient Intelligence. Lecture Notes in Computer Science, vol. 7657. pp. 470–477. Springer Berlin Heidelberg (2012).

Tomsen, V.: Response Time of a Thermometer. Phys. Teach. 36, 540–541 (1998).

Touloukian, Y.S., Kirby, R.K., Taylor, R.E., Desai, P.D.: Thermal Expansion - Metallic Elements and Alloys. In: Touloukian, Y.S., Kirby, R.K., Taylor, R.E., and Desai, P.D. (eds.) The TPRC Data Series, vol 12. Plenum Publishing Corporation (1975).

Wagman, D.D., Evans, W.H., Parker, V.B., Schumm, R.H., Halo, I., Bailey, S.M., Churney, K.L., Nuttall, R.L.: The NBS Tables of Chemical Thermodynamic Properties. J. Phys. Chem. Data. 11, Supp., 1–392 (1982).

Yacobi, B.G.: Semiconductor Materials : An Introduction to Basic Principles. Kluwer Academic Publishers, Secaucus, NJ (2002).

The RTD. Omega Engineering Inc., http://www.omega.com/temperature/Z/TheRTD.html (2014) Accessed 4 May 2014.

ANSI and IEC Color Codes for Thermocouples, Wire and Connectors/Thermocouple Tolerances. Omega Engineering Inc., http://www.omega.com/temperature/pdf/tc_colorcodes.pdf (2014) Accessed 05 May 2014.

Electrical Conductivity and Resistivity. NTD Resource Center., http://www.ndt-ed.org/Education Resources/CommunityCollege/Materials/Physical_Chemical/Electrical.htm (2014) Accessed 04 June 2014.

Physical Properties of Thermoelement Material. Omega Engineering Inc., http://www.omega.com/temperature/Z/pdf/z016.pdf (2014) Accessed 04 May 2014.

Chapter 3
Sensor Measurement Capability

Gopi Krishnan

Sensor accuracy, repeatability, and reproducibility are some of the key metrics that govern the usefulness of the sensor and the associated metrology system used for measuring a physical parameter. Sensor measurement capability determines which application a given sensor can be used for. This chapter describes the definition of different terminologies associated with a sensor and metrology variability—accuracy, precision, repeatability, reproducibility, linearity, and stability. The chapter also describes the methodology of evaluating measurement system capability.

3.1 Introduction

The temperatures at multiple locations in an electronic package assembly are commonly measured parameters (Fig. 3.1). The locations include

1. the die or junction (T_j),
2. the heat spreader or case (T_c),
3. the heat sink (T_s—sink),
4. and the ambient environment (T_a—ambient).

We should note here that while there are numerous measurement methods or metrologies that may be employed at the different package locations, RTD's (Resistance Temperature Detector's) are commonly used for die level measurement, with thermocouples used at the heat spreader, heat sink, and ambient level. In this chapter we will continue to use this premise.

G. Krishnan (✉)
Intel Corporation, Santa Clara, USA
e-mail: gopi.krishnan@intel.com

© Springer Science+Business Media New York 2015
C.M. Jha (ed.), *Thermal Sensors*, DOI 10.1007/978-1-4939-2581-0_3

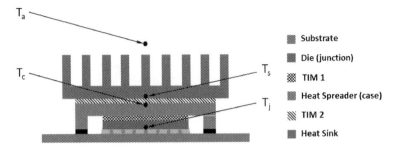

Fig. 3.1 A flip chip package with a heat spreader and heat sink. Temperature measurement locations include the junction (T_j), case (T_c), heat sink (T_s), and ambient environment (T_a)

Now, the measured data is used either directly or to calculate one among several metrics such as the junction to case thermal resistance (R_{jc}) as may be defined by Eq. 3.1:

$$R_{jc}(°C - cm^2/W) = \frac{(T_j - T_c)A}{P} \tag{3.1}$$

where P (W) is dissipated power, and A (cm^2) is the area of the die.

In turn, the direct or indirect data is used for:

1. package design validation,
2. process characterization and development, and
3. product certification and disposition.

Example 1.1 A new TIM1 (Thermal interface material) is being evaluated for use on an electronic package. With the new TIM1, the package dissipates 50 W, with the die and case temperature measuring 100 and 90 °C respectively. The die area is 2 cm^2. Determine if the package with the new TIM meets the thermal resistance (R_{jc}) requirement of 0.5 °C-cm^2/W?

Solution

Known: P = 50 W, T_j = 100 °C, T_c = 90 °C, A = 2 cm^2, R_{jc} Target = 0.5 °C-cm^2/W

Find: R_{jc} of package with new TIM1

Analysis: The package R_{jc} is calculated using Eq. 3.1

$$R_{jc} = \frac{(T_j - T_c)A}{P} = \frac{(100\,°C - 90\,°C)2\,cm^2}{50\,W} = 0.4\,°C - cm^2/W$$

The package with the new TIM1 meets the thermal requirement.

Given the significance of the decisions being made using the temperature data, it is vital to assess if the metrology is capable of meeting the needs of the decision making process by quantifying the inherent variation the metrology introduces into a measurement. The following section provides an overview of the measurement capability analysis on temperature sensors.

3.2 Variability in Measurements

Whenever a temperature measurement system is used to measure or observe the true value of temperature, a measurement error is introduced by the system to varying amounts, and may be denoted as:

$$Observed\ value = True\ value + Measurement\ error \qquad (3.2)$$

Measurement system errors arise from a multitude of sources including those from the:

1. Environment: Temperature oscillations, vibration
2. Procedure: Calibration, handling and installation, data recording and calculation
3. Equipment: sensor manufacturing, hardware and data acquisition system, sensor degradation (Fig. 3.2).

3.3 Accuracy

Accuracy is the difference between the true value and the average value of measurements. Bias is often used to enumerate accuracy as

$$\text{Bias} = \mu_{measurement} - \mu_{true}$$

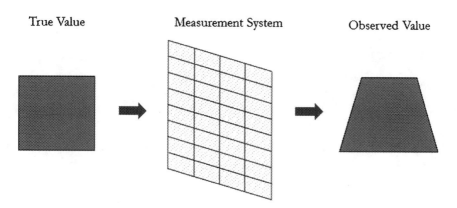

Fig. 3.2 The measurement error introduced by a measurement system when observing the true value of a parameter

Example 1.2 Calculate the bias and standard deviation of the temperature measurements from two RTD's located on a die that is independently ascertained to be at 80 °C.

Measurements (°C)

RTD 1: 99.1, 97.0, 93.5, 96.5, 98.5, 92.5
RTD 2: 71.9, 75.9, 65.4, 72.9, 65.7, 71.7

Solution

Known: RTD observations (see list above), True Value = 80 °C
Find: Bias and Standard Deviation
Analysis: The average of the observations is calculated by

$$\mu = \frac{\sum_{i=1}^{N} T_i}{N}$$

and the standard deviation using:

$$\sigma = \sqrt{\frac{\sum_{i=1}^{N} (T_i - \mu)^2}{N - 1}}$$

Executing the calculation, the bias and standard deviation for each RTD are tabulated.

Observations		
Reading	RTD 1	RTD 2
1	99.1	71.9
2	97.0	75.9
3	93.5	65.4
4	96.5	72.9
5	98.5	65.7
6	92.5	71.7
Summary statistics		
Average	96.5	71.5
Std dev.	2.3	4.4
Bias	16.5	−8.5

The bias in RTD 1 is positive and in RTD 2 negative.

Example 1.3 Consider four arrows hitting a target, with the center being the goal. In each of the cases, comment on the accuracy and precision of the result.

Solution

Known: Locations of the arrows relative to each other and the goal.
Find: Qualitatively describe the accuracy and precision.
Analysis:

	Target			
	1	2	3	4
Accurate	No	No	Yes	Yes
Precise	No	Yes	No	Yes

Example 1.4 Four RTD's take four temperature measurements each, on a die which is independently measured to be at 100 °C. Which of the RTD's have the highest accuracy, highest precision, and which RTD is more likely to be preferred for future use?

Measurements (°C)

RTD 1: 106, 93, 98, 112
RTD 2: 104, 103, 101, 103
RTD 3: 101, 100, 99, 98
RTD 4: 97, 85, 114, 102

Solution

Known: RTD observations (see list above), True Value = 100 °C
Find: A measure of accuracy and precision for each RTD
Analysis: The average, standard deviation, and error (True-Average) of each RTD is calculated and listed below

Observations				
Reading	RTD 1	RTD 2	RTD 3	RTD 4
1	106	104	101	97
2	93	103	100	85
3	98	101	99	114
4	112	103	98	102
Summary statistics				
Average	102.3	102.8	99.5	99.5
Std dev.	8.4	1.3	1.3	12.0
Error	−2.3	−2.8	0.5	0.5

(continued)

(continued)

Observations				
Reading	RTD 1	RTD 2	RTD 3	RTD 4
Assessment				
Accuracy	Low	Low	High	High
Precision	Low	High	High	Low

The most accurate and precise RTD's are ones that exhibit the least error and standard deviation respectively. Thus, RTD's 3 and 4 exhibit the highest accuracy while RTD's 2 and 3 show the highest precision. RTD 3 displays relatively the highest accuracy and precision and is more likely to be used.

A complete accuracy study of an RTD system may be conducted as follows

1. Select the measurement parameter of interest e.g. Temperature from RTD.
2. Determine the 'true value' of the parameter, obtained by an independent metrology of higher standard (e.g. Calibrated accurate reference RTD).
3. Measure the temperature on a single part 16 times. Between each measurement, remove the part from the measurement system (e.g. socket), reseat it, and then allowed the part to reach steady state.
4. By performing the measurements using a single operator and within a short time span, we minimize all other sources of noise other than that associated with the metrology and the operational method.
5. Plotting a data in run order reveals any trends or outliers.
6. Calculate the mean and standard deviation of the data set, and perform a statistical test to compare the mean of the measurements to the reference temperature to determine if the bias is statistically significant or not.
7. If the bias is deemed to be significant, calibration of the RTD is required upon which steps 3 through 6 are repeated.
8. If the bias is statistically insignificant then no calibration or correction of the metrology is required.

3.4 Precision

Precision of a measurement system consists of its repeatability and reproducibility and may be estimated as

$$\sigma^2_{measurement\ system} = \sigma^2_{repeatability} + \sigma^2_{reproducability}$$

Example 1.5 The repeatability and reproducibility of a RTD measurement system is estimated to be 0.5 °C ($\pm 1\sigma$) and 0.9 °C ($\pm 1\sigma$) respectively. Calculate the precision of the measurement system.

Solution

Known: $\sigma_{repeatability} = 0.5$, $\sigma_{reproducibility} = 0.9$
Find: Precision

Analysis: The precision or total measurement system variability is

$$\sigma^2_{\text{measurement system (ms)}} = \sigma^2_{\text{repeatability}} + \sigma^2_{\text{reproducability}}$$

$$\sigma^2_{\text{ms}} = 0.5^2 + 0.9^2 = 1.06$$

Therefore the precision is $\sqrt{\sigma_{ms}}$ or 1.03 °C.

3.5 Repeatability

Repeatability is the variation in the repeated measurement of single component under same operational conditions such as time of day, operator and local conditions. Static repeatability measures the variability of the measurement system solely, while dynamic repeatability measures the variability of both the measurement system and measurement method.

Example 1.6 The temperature of die on an electronic package with an embedded RTD is repeatedly measured 10 times, where after each measurement the package is removed and then placed back in the measurement system socket. Calculate the dynamic repeatability of the system.

Measurements (°C): 87.3, 86.1, 87.0, 86.9, 87.7, 86.4, 87.1, 86.0, 86.1, 86.1

Solution

Known: 10 measurements
Find: Dynamic repeatability
Analysis: Calculating the standard deviation of the measurements using:

$$\sigma_{rpt} = \sqrt{\frac{\sum_{i=1}^{N}(T_i - \mu)^2}{N-1}}$$

yields $\sigma^2_{\text{repeatability}} = 0.6$ °C.

3.6 Reproducibility

Reproducibility is the cumulative measurement variation that arises due to different conditions that may exist during the measurement.

The following steps may be used to conduct a reproducibility study

1. Experimental design: To capture all the sources of measurement variation, an experiment needs to incorporate multiple noise factors set at appropriate levels. In addition, more than one part needs to be subject to the sources of measurement variation, to uncover any metrology/part interaction. The table below is

an example of experimental conditions that may be relevant when conducting a reproducibility study on the RTD measurement system.

Factors/levels	
Factor	Levels
RTD	2
Operator	3
Day of week	3
Repeated measurement	3

And experiment that permits qualifying the effect of the factors and their interactions (e.g. a full factorial) is designed.

2. Execution: Ideally the experiment should be run in a random order, however, if logistics prevent this, efforts to minimize run order effects must be made.
3. Data Analysis: With the experiment executed, the data collected can be analyzed to quantify the reproducibility.

Example 1.7 A RTD is embedded in multiple electronic packages to help evaluate the thermal performance of thermal interface materials. To determine the measurement uncertainty of the temperature measurement, a reproducibility study is conducted considering the day of week, different operator and multiple packages. The variances associated with variation in day and operators are estimated to be 0.7 and 1.2 respectively. The variance in repeatability was assessed as 0.2. Estimate the precision of the RTD measurement.

Solution

Known: $\sigma^2_{day} = 0.7$, $\sigma^2_{operator} = 1.2$, $\sigma^2_{package} = 0.4$, $\sigma^2_{repeatability} = 0.2$
Find: Precision (σ_{ms})
Analysis: The variation from one unit to another unit (in other words one sample to another sample) is not considered in the evaluation of the reproducibility. The precision or total measurement system variability is

$$\sigma^2_{measurement\ system\ (ms)} = \sigma^2_{repeatability} + \sigma^2_{reproducability}$$

where

$$\sigma^2_{reproducability} = \sigma^2_{day} + \sigma^2_{operator}$$

The precision is calculated as $\sqrt{\sigma_{ms}} = \sqrt{0.2 + 0.7 + 1.2} = 1.4\ °C$.

3.7 Linearity

Linearity is a measure of the consistency of the Bias and precision of the measurements over the entire range of the system (Fig. 3.3).

A non-linearity in bias results in challenges in calibration, while a non-linearity in precision results in changing variability at each point.

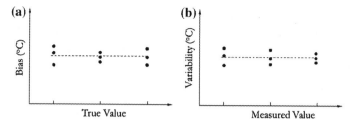

Fig. 3.3 **a** Bias is consistent across the range of true values. **b** The variability is consistent at each measured value

3.8 Stability

A temperature measurement system that exhibit steady and predictable measurements over time both in the bias (mean) and precision (standard deviation) is considered stable. A process that is stable does not show sudden shifts, outliers, trends or non-random behavior and may be evaluated for stability using control charts. On a control chart a stable measurement system will reveal purely random behavior with the absence of variation due to special causes.

Example 1.8 Evaluate the stability of each of the following temperature measurement systems.

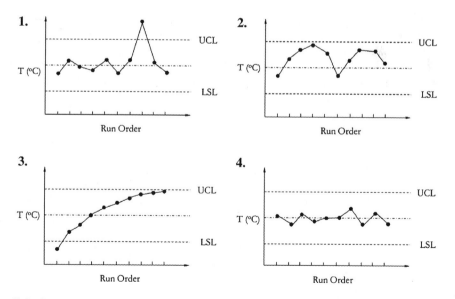

Solution

Known: Control Charts with LCL (lower control limit), LSL (lower specification limit), UCL (upper control limit), and USL (upper specification limit) are specified.

Find: Assess stability of measurement system

Analysis: Systems 1, 2, and 3 are not stable due to the presence of an outlier, cyclic behavior, and a trend, respectively. System 4 is stable due to the absence of any special causes or non-random behavior.

3.9 Measurement System Capability

Thermal metrologies are often used to disposition units based on process specification limits. The capability of a temperature metrology to perform this task lies in the extent of the measurement variation relative to the process specification range. Figure 3.4 shows a process with two-sided specifications defined by a LSL (Lower Spec Limit) and the USL (Upper Spec Limit), where the total variation in the measurement takes up the specification window (Fig. 3.4).

The measurement capability is often expressed by the Precision to Tolerance (P/T) Ratio, which articulates the fraction of the process specification window that is enveloped by measurement variation. For a two-sided process specifications it is defined as

$$P/_T = \frac{6\sigma_{ms}}{USL - LSL}$$

and for a one-sided process specification as

$$P/_T = \frac{3\sigma_{ms}}{|USL \text{ or } LSL - Process\ Mean|}$$

A smaller P/T ratio results in less of the process specification window being shrouded by the measurement variation.

Customarily a measurement system with a P/T ratio less than 0.3 is deemed to be capable. A system exceeding 0.3 is considered not capable as a result of its low precision.

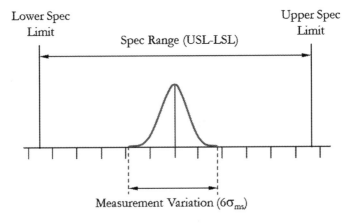

Fig. 3.4 The measurement variation enveloping part of the process specification range

Example 1.9 The specification window for a particular thermal driven process is stated as 150 ± 4 °C. Determine if a temperature measurement system with a variation captured by $\sigma_{ms} = 0.3$ °C is considered capable of dispositioning units in this process specification window.

Solution

Known: LSL $= 146$ °C, USL $= 154$ °C, $\sigma_{ms} = 0.3$ °C
Find: P/T ratio
Analysis: The system P/T ratio is calculated as:

$$P/T = \frac{6\sigma_{ms}}{USL - LSL} = \frac{6\,(0.3\,°C)}{154\,°C - 146\,°C} = \frac{1.8\,°C}{8\,°C} = 0.23$$

The P/T ratio is less than 0.3 indicating that the system is capable of dispositioning units in this process window.

In addition to dispositioning units, thermal measurement systems are used to perceive changes in a process. In the case where no process specification limits are available, an alternative metric to assess metrology capability is the Signal to Noise Ratio (SNR) of the measurement system, defined as

$$\text{SNR} = \frac{\sigma_{process}}{\sigma_{ms}}$$

Conventionally a SNR > 3 is used to indicate a capable measurement system.

Example 1.10 The variability of temperature in a thermal driven process is assessed to be $\sigma_{process} = 0.7$ °C, with no specification limits set. Determine if a temperature measurement system with a $\sigma_{ms} = 0.3$ °C is considered capable of discerning changes in this process.

Solution

Known: $\sigma_{process} = 0.7$ °C, $\sigma_{ms} = 0.3$ °C
Find: P/T ratio
Analysis: The measurement system SNR ratio is calculated as:

$$P/T = \frac{\sigma_{process}}{\sigma_{ms}} = \frac{0.7\,°C}{0.3\,°C} = \frac{0.7\,°C}{0.3\,°C} = 2.3$$

The SNR ratio is less than 3 indicating that the system is not capable of capturing changes in the process. The measurement system needs to be improved by identifying and reducing the largest sources of variation that may be regulated.

Example 1.11 Perform a measurement capability study on a temperature measurement system to assess accuracy and repeatability.

Solution

Find: Accuracy and repeatability
Analysis:

a. Accuracy
    The temperature from an RTD embedded in a die in a thermal test vehicle is
    measured 16 times. Between each measurement, the part is removed and re-
    socketed, with the test vehicle allowed to reach steady state. A single tech-
    nician performs the test within a day on the same socket. Using a higher
    standard reference temperature probe, the die temperature is determined to be
    70 °C at steady state.
    The results of the measurements are as follows:

    68.15, 71.92, 71.98, 68.19, 70.30, 72.14, 70.11, 72.96, 72.75, 68.84, 72.00,
    72.54, 68.95, 71.22, 71.95, 68.98

    A run order plot (Fig. 3.5) of the measurements reveals no trends or outli-
    ers. The mean and standard deviation of the measurements are 70.81 and
    1.71 respectively with a resulting **bias of 0.81C**. A statistical test compar-
    ing the mean of the measurements to the reference temperature shows that
    the bias is not statistically significant, suggesting that no recalibration is
    required (Fig. 3.6).

b. Repeatability
    The temperature of the same test vehicle is now measured 30 times, where
    once again between each measurement the part is removed and re-socketed,
    with the test vehicle allowed to reach steady state. A single technician per-
    forms the test within a day on the same socket. Assume that the RTD will be
    used to disposition units within a process window of 70 ± 12 °C.

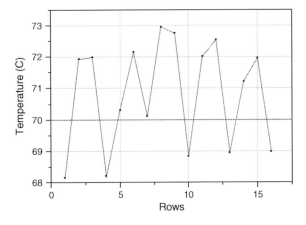

Fig. 3.5 A run order plot of the temperature measurements

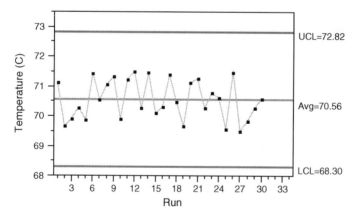

Test Mean

Hypothesized Value	70
Actual Estimate	70.8114
DF	15
Std Dev	1.71329

t Test

Test Statistic	1.8943
Prob > |t|	0.0776
Prob > t	0.0388*
Prob < t	0.9612

Fig. 3.6 Statistical test of the mean of the measurement

Fig. 3.7 A control chart to assess stability

The results of the measurements are as follows:

71.12, 69.66, 69.92, 70.28, 69.87, 71.44, 70.54, 71.05, 71.33, 69.91, 71.23, 71.49, 70.28, 71.45, 70.11, 70.31, 71.38, 70.48, 69.66, 71.12, 71.26, 70.27, 70.77, 70.62, 69.58, 71.48, 69.51, 69.83, 70.27. 70.59

Plotting the measurements in run order does not reveal any trends or outliers. When plot against the calculated control limits, no out-of-control points are observed with the data deemed stable (Fig. 3.7).

The mean and standard deviation of the measurements are 70.55 and 0.66 respectively. The P/T ratio is then calculated as

$$P/T = \frac{6\sigma_{ms}}{USL - LSL} = \frac{6(0.66\,^\circ C)}{82\,^\circ C - 58\,^\circ C} = \frac{3.96}{24} = 0.17$$

With a P/T repeatability ratio less that 0.2 suggesting that the system is capable for repeatability.

Chapter 4
Microprocessor Temperature Sensing and Thermal Management

Chandra Mohan Jha and Jaime A. Sanchez

Microprocessors, while operating, exhibit non uniform temperature distribution and are rated by their maximum junction temperature. The thermal design and the cooling requirements of the microprocessor package govern the thermal specification of the product. Accurate evaluation of the junction temperature, T_j, is important in defining the product thermal specifications. It requires a combination of modeling and experimental approaches to estimate the T_j of a microprocessor product. It is an iterative process between the T_j estimate and the thermal design of the product. If the estimate of T_j exceeds the requirements of the specification, the thermal design and the cooling solution of the product needs to be modified. A temperature sensor within the silicon die measures the real time T_j during the operation of the product. The T_j measurement ensures that the product is working below the threshold limit as defined in the specification. An error in the T_j estimates and the T_j measurements can result in negative thermal impact of the product thereby reducing either its performance or operating life. Understanding the overall thermal management and the error analysis helps in effective temperature control of the junction temperature. This chapter describes the T_j evaluation process, temperature sensing methods, the thermal management of the microprocessor, and the sensitivity analysis showing the thermal impacts.

C.M. Jha (✉) · J.A. Sanchez
Intel Corporation, Santa Clara, USA
e-mail: cmjha75@gmail.com; chandra.mohan.m.jha@intel.com

J.A. Sanchez
e-mail: Jaime.a.sanchez@intel.com

© Springer Science+Business Media New York 2015
C.M. Jha (ed.), *Thermal Sensors*, DOI 10.1007/978-1-4939-2581-0_4

4.1 Junction Temperature and Thermal Sensors

Microprocessors have a complex architecture (Moore 1965; Borkar 1999; Yuffe et al. 2011; Borkar 2011) with CPUs (central processing units) and GPUs (graphics processing units) that operate up to several gigahertz (GHz) of frequency. The switching of transistors and their intrinsic characteristics result in power dissipation. The power source in a microprocessor is non uniform and depends on various factors which will be explained later in the chapter. Power distribution of a microprocessor for a given software application, called workload, is shown in Fig. 4.1. Different types of workloads produce different power distribution. Figure 4.1a shows a workload that result into concentrated power inside the microprocessor die whereas Fig. 4.1b shows a workload with relatively distributed power. In both cases, the power is highly non-uniform. The temperature distribution corresponding to the non-uniform power and its comparison with that of the uniform power is shown in Fig. 4.2.

The microprocessor silicon die is assembled in a package with thermal cooling solutions attached on top of it as shown in Fig. 4.3. The package is mounted on the motherboard which in turn is used in a computer system. A typical package constitutes substrate, die, and integrated heat spreader (IHS). A thermal cooling solution such as a heat sink is attached on top of the package to cool the microprocessor. The die and the IHS interface has a thermally conductive material called thermal interface material (TIM). To distinguish it from other interfaces, it is commonly referred to as TIM1. Similarly, the IHS and the heat sink interface has another thermal interface material called TIM2.

When the die is powered on, it heats up and has a certain temperature distribution. The maximum temperature in the microprocessor die is called junction

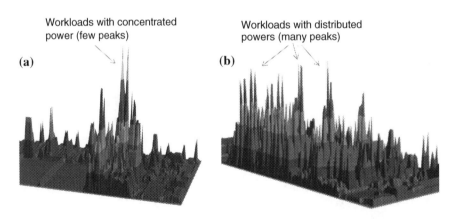

Fig. 4.1 Normalized non-uniform power maps in a typical microprocessor for two different types of workloads. The height of the peak signifies the magnitude of the power dissipation (W). **a** Powermap corresponding to workloads with concentrated power in the die. **b** Powermap corresponding to workloads with distributed power in the die

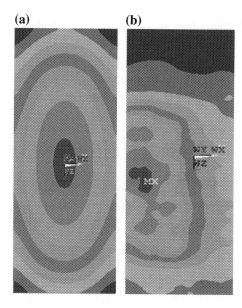

(a)   **(b)**

Fig. 4.2 Temperature distributions in microprocessor die for **a** uniform power and **b** non-uniform power

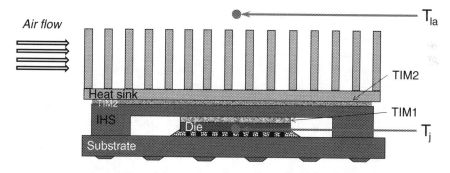

Fig. 4.3 Schematic of a typical microprocessor package and the cooling system stack up—die, thermal interface material between die and integrated heat spreader (TIM1), integrated heat spreader (IHS), second thermal interface material between IHS and heat sink (TIM2), and heat sink. T_j is the die junction temperature and T_{la} is the ambient temperature

temperature T_j. The ambient temperature nearby the heat sink is referred as local ambient T_{la} of the microprocessor heat sink system. There is a thermal resistance between the microprocessor junction and the local ambient which depends on the design of the package and the cooling system. It is the combination of the micro-processor power and the junction to ambient thermal resistance that governs the value of T_j.

4.1.1 Thermal Design Power

When the microprocessor product is operating in the computing system, it is designed so that the maximum temperature of the die, or the junction temperature T_j, is always maintained at or less than a certain limit. This limit is imposed by the thermal designers to maintain the product's reliability during its specified operating life. The performance of the microprocessor depends on the temperature at which it operates. Higher operating temperature can cause performance degradation over a relatively short period of time. Hence, to maintain the reliable operation of the product, a threshold limit on the maximum junction temperature is used, called $T_{j\text{-}max}$. If the T_j exceeds the $T_{j\text{-}max}$ the microprocessor is forced to reduce its input power. This process is called throttling. The maximum sustained microprocessor power corresponding to $T_{j\text{-}max}$ is called thermal design power (TDP) and is predominantly dependent on the dynamic power and the leakage power of the microprocessor (Borkar 2011) given by Eq. (4.1).

$$TDP = P_{dynamic} + P_{leakage} \qquad (4.1)$$

$P_{dynamic}$ is the dynamic power and is given as

$$P_{dynamic} = C_{dyn} V_{cc}^2 f \qquad (4.2)$$

where,

C_{dyn} is a capacitance and depends on the manufacturing parameters
V_{cc} is the supply voltage
f is microprocessor frequency

From Eq. (4.2), microprocessors with higher frequency will have higher dynamic powers causing localized hot spots as shown in the Fig. 4.2b. The dynamic power also depends on the type of the workload used in the microprocessor. The TDP acts as a thermal specification of the product and is defined for a specific workload at a given frequency.

The leakage power is due to the transistor leakage currents. It depends on the fabrication technology, processes and other physical properties. It also depends on the temperature of the device. Higher die temperature results in higher leakage power. Leakage power does not contribute to the computing performance of the microprocessor. However, it adds significant challenges in its thermal management.

Since TDP is the maximum power corresponding to $T_{j\text{-}max}$, Eq. (4.1) can be written as:

$$TDP = P_{dynamic} + P_{leakage} = \frac{T_{j-max} - T_{la}}{\psi_{ja}} \qquad (4.3)$$

T_{la} is the local ambient temperature of the overall system as shown in Fig. 4.3 and $\psi_{ja}(°C/W)$ is the thermal resistance of the cooling system between the die junction to the local ambient.

Microprocessors do not operate at TDP or $T_{j\text{-}max}$ all the time. In fact, most of the software applications result in the operating power less than the TDP, even though the entire system is designed for TDP. When the total power of the microprocessor (P_{total}) is less than the TDP, Eq. (4.3) can be modified as:

$$P_{total} = P_{dynamic} + P_{leakage} = \frac{T_j - T_{la}}{\psi_{ja}} \tag{4.4}$$

or,

$$T_j = P_{total} \times \psi_{ja} + T_{la} \tag{4.5}$$

The CPU needs to meet its thermal specification in terms of the TDP without exceeding $T_{j\text{-}max}$. For example, if ψ_{ja} happens to deteriorate and exceed beyond its designed value due to some unknown reasons, the T_j will reach $T_{j\text{-}max}$ even if the P_{total} is significantly less than the TDP, violating the thermal specification. Equation (4.5) shows that T_j depends on the total power, thermal resistance between junction to ambient, and the ambient temperature. Any change in these parameters results in changes to T_j. For proper and reliable operation of the device, T_j needs to be continuously monitored and ensured that it does not exceed $T_{j\text{-}max}$.

Example 1 A microprocessor is dissipating 100 W. The junction to ambient thermal resistance is 0.5 °C/W. What is the junction temperature, T_j? Assume 30 °C ambient temperature. If the $T_{j\text{-}max}$ is 100 °C, what is the TDP?

Solution
From Eq. (4.5), T_j is given as

$$T_j = P_{total} \times \psi_{ja} + T_{la}$$

Known parameters:

$$P_{total} = 100\,\text{W}$$
$$\psi_{ja} = 0.5\,°\text{C/W}$$
$$T_{la} = 30\,°\text{C}$$

Therefore,

$$T_j = 100 \times 0.5 + 30 = 80\,°\text{C}$$

This means with 100 W input power, the junction temperature of the given microprocessor with the junction to ambient thermal resistance of 0.5 °C/W, is 80 °C. If the $T_{j\text{-}max}$ is higher than 80 °C, then the TDP should be higher than 100 W and can be computed using Eq. (4.3) as shown below.

$$TDP = \frac{T_{j\text{-}max} - T_{la}}{\psi_{ja}}$$

Known parameters:

$$T_{j\text{-}max} = 100\,°\text{C}$$
$$\psi_{ja} = 0.5\,°\text{C/W}$$
$$T_{la} = 30\,°\text{C}$$

Therefore,

$$\text{TDP} = \frac{100 - 30}{0.5} = 140\,\text{W}$$

4.1.2 Measurement of T_j

The T_j is measured real time during the microprocessor operation. Multiple temperature sensors are used for the measurement, as shown in Fig. 4.4. The objective is to measure the maximum temperature as close to the die hot spots as possible. However, it is not feasible to have a temperature sensor that can directly measure the maximum hot spot temperature in the die. There are several reasons for this:

1. Complex architecture: The microprocessor architecture design generally has limited real estate die space (premium space) for the temperature sensor in the immediate vicinity of the CPU cores where there is high probability of having the hot spots.
2. Varying workloads: The hot spot location varies with different workloads as shown in Fig. 4.5. When a computer runs a highly intensive computing program it will have a different hot spot location, compared to the scenario when it runs a program that is not as compute intensive.
3. Proximity to the hot spot: The hot spot generally occurs at one of the CPU cores with high power density and there has to be some physical separation between the temperature sensor and the CPU core.

The difference in the maximum temperature and the temperature measured by the sensor has to be accounted for. This ΔT can be computed with the help of 3D modeling where the temperature distribution across the die is generated. The difference in temperatures at the locations of the sensors and the hot spots from the model gives ΔT for every sensor.

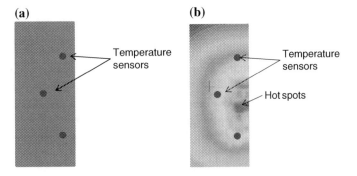

Fig. 4.4 a Microprocessor silicon die with temperature sensors hypothetically shown at different locations, **b** die with non-uniform temperature distribution showing the difference in the locations of the hot spot and the temperature sensors

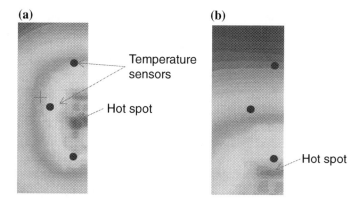

Fig. 4.5 Die temperature distribution with different workloads. The location of hot spot varies with the work loads. **a** Hot spot located at the *edge* of the die. **b** Hot spot located at the *corner* of the die

4.1.2.1 CMOS Sensors in Microprocessors

The inbuilt die temperature sensors are compatible with the CMOS (complementary metal oxide semiconductor) fabrication process used for the manufacturing of microprocessor. Resistance temperature detectors, RTDs and diode thermal sensors are CMOS compatible and commonly used for the T_j measurement. The output signal of the sensor can be either analog or digital. The sensor generally used in the microprocessor is called a digital temperature sensor (DTS). It is an analog diode thermal sensor with an analog-to-digital converter (A/D) circuit built in as a part of the sensor. Figure 4.6 shows the block diagram of a DTS. The output of the analog sensor is converted into the digital signal by using the A/D converter. The digital output is used for the thermal control of the microprocessor. Depending upon the types of the product and the applications, different control monitors can be employed. For example, a trigger can be set when the T_j reaches some kind of threshold limits—Specification limit, Throttling limit, and Thermtrip limit.

1. Specification limit: It is the T_{j-max} defined in the thermal specification of the microprocessor.
2. Throttling limit: It can be set either below or equal to the spec limit depending upon the algorithm designed by the system designer. The objective of the throttling limit is to take pre-emptive action to prevent T_j from exceeding the T_{j-max}.

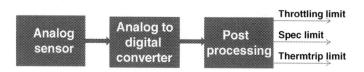

Fig. 4.6 Block diagram of a digital temperature sensor (DTS)

3. Thermtrip limit: It is a limit when a catastrophic shutdown of the microprocessor and the system takes place. It can be a case of thermal runaway due to some system malfunction or cooling issues.

There are many challenges associated with the accuracy of the DTS output:

1. Analog sensor noise: The diode sensor has inherent noise and bias which limits the minimum temperature precision and accuracy that can be achieved. In addition, there is sensor-to-sensor variability due to the manufacturing process variation from one part to another.
2. A/D circuit inaccuracy: Analog-to-digital circuitry coupled with the manufacturing process variation induces additional noise and variation in the DTS output.
3. Sensor calibration: A limited sample of sensors is calibrated in a controlled isolation bath chamber. Error associated with the sampling and the calibration changes the final sensor accuracy.
4. Testing: All parts are tested in a certain test environment at a given set temperature. Variations in the test parameters and the set temperature affect the DTS accuracy.

4.1.2.2 Diode Sensors

Temperature sensors used in microprocessors are usually thermal diodes. In an ideal diode, the forward current through the PN junction can be expressed by Eq. (4.6a) which is known as the Shockley diode equation.

$$i = i_s(e^{\frac{ve}{nkT}} - 1) \qquad (4.6a)$$

where,

i_s is the saturation current
v is the voltage drop across the junction
e is the electron charge
T is the temperature in Kelvin
k is the Boltzmann's constant ($1.3806488 \times 10^{-23}$ m^2 kg s^{-2} K^{-1})
n is the ideality factor and depends on the manufacturing process technology. Under ideal scenario $n = 1$

Since the saturation current is a property of the circuit dependent on the characteristics of the silicon fabrication process it can be considered a constant. This means that Eq. (4.6a) has three main variables: the current across the thermal diode (i), its temperature (T) and the voltage across the diode (V). In Fig. 4.7, Eq. (4.6a) is plotted to understand the influence of various terms with the x-axis showing the voltage drop across the junction normalized to the thermal voltage (nkT/e) and the y-axis showing the current across the thermal diode normalized to the saturation current.

Equation (4.6a) is usually re-written considering that it is dominated by the exponential term, giving Eq. (4.6b)

$$i = i_s e^{\frac{V_{BE}}{nkT/q}} \qquad (4.6b)$$

Fig. 4.7 Graphical representation of the diode current i for a Shockley diode

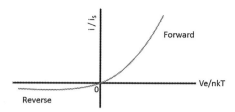

A common operating mode for thermal diode sensors is in a dual-current mode, where two source currents are used at different magnitudes which result in two back-emitter voltages being generated. Let $i_1 = i_s e^{\frac{V_{BE1}}{nkT/q}}$ and $i_2 = i_s e^{\frac{V_{BE2}}{nkT/q}}$ then one can re-arrange the terms to generate Eq. (4.6c) which relates temperature T to the back-emitter voltages 1 and 2, as well as the source currents 1 and 2 as a function of the ideality factor n which in turn depends of the silicon process.

$$\Delta V_{BE} = V_{BE1} - V_{BE2} = \frac{nkT}{q} \ln\left(\frac{I_1}{I_2}\right) \tag{4.6c}$$

In order to use the thermal diode to measure the temperature of the microprocessor, one must then be able to source a known current and measure the voltage across the diode.

4.1.2.3 Diode/DTS Calibration

DTS calibration is a key step in the manufacturing process of microprocessors. These sensors are used to measure the temperature of the device during operation to determine the temperature margin with respect to $T_{j\text{-}max}$. This information is used by the microprocessor to lower its frequency if it reaches $T_{j\text{-}max}$, which lowers the power and thus the temperature. This mechanism is called throttling and is part of the power management design of the device. The temperature information can also be used to increase the frequency of the device in Turbo mode to improve its performance under certain applications based on temperature margin with respect to $T_{j\text{-}max}$. Finally, for mobile products that operate on battery, the temperature can be used to calculate the leakage component of the power and thus effectively manage the power consumption from a battery.

Typical microelectronics have temperature accuracy specification on the DTS of ± 5 °C (source: http://www.cypress.com/?docID=49440). In order to meet this specification, the DTS usually need two points of calibration: at a low temperature and at a high temperature ($T_{j\text{-}max}$). Under ideal conditions, a bath chamber is used to calibrate temperature sensors (see the next section).

4.1.2.4 Resistance Temperature Detectors (RTDs)

The RTD is commonly designed as a serpentine resistive structure. The accuracy of the RTD depends on the accuracy of the resistance measurement. It is designed to minimize the joule heating loss during the measurement of the resistance of the RTD. The joule heat can alter the resistance of the RTD causing error in the temperature measurement. The measurement setup and the wire resistance of the instrument can also cause error in the resistance measurement. The wire resistance error can be avoided by using Kelvin resistance measurement, also called 4-wire resistance measurement method.

4-wire Measurement: The 4-wire method is used to accurately measure the resistance in the circuit and is often used for the temperature sensor measurement. A non-4-wire method (or conventional 2-wire method) of measuring the resistance includes all the resistances of traces and wires in the circuit loop which are not part of the heater resistance thereby inducing error in the measurement. In the 4-wire method, the resistance contribution from the traces or wires which are not part of the heater resistance is minimized thereby reducing the error. Figure 4.8 shows the comparison of 2-wire and 4-wire methods.

The 2-wire method has 2 terminals connected to the resistor whereas 4-wire method has 4 terminals. The data acquisition system captures the resistance of not only the resistor but also the wires or traces connected to the resistor. By keeping

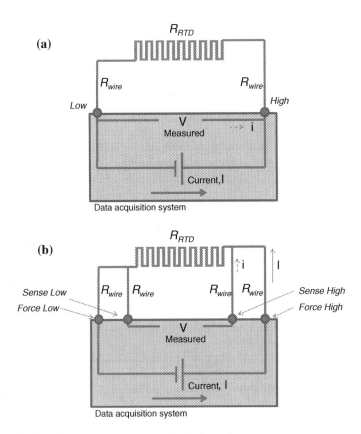

Fig. 4.8 **a** 2-wire resistance measurement and **b** 4-wire resistance measurement

the voltage connection as close to the resistor, the wire resistance can be eliminated. Equation (4.7) shows the measured voltage in the 2-wire case.

$$V_{2\text{-}wire} = (I + i)(R_{RTD} + R_{wires}) \tag{4.7}$$

where,

I is the forcing current for the measurement
i is the sensing current which is much smaller than the forcing current

Assuming $i \ll I$, Eq. (4.7) can be simplified as shown below.

$$V_{2\text{-}wire} = I(R_{RTD} + R_{wires}) \tag{4.8}$$

or,

$$V_{2\text{-}wire} = IR_{measured\text{-}2wire} \tag{4.9}$$

or,

$$R_{measured\text{-}2wire} = \frac{V_{2\text{-}wire}}{I} = R_{RTD} + R_{wires} \tag{4.10}$$

Similarly, for the 4-wire case the voltage output is given by Eq. (4.11).

$$V_{4\text{-}wire} = (I + i)(R_{RTD}) + iR_{wires} \tag{4.11}$$

Since $i \ll I$, Eq. (4.11) can simplified as shown below.

$$V_{4\text{-}wire} = IR_{RTD} \tag{4.12}$$

or,

$$R_{measured\text{-}4wire} = \frac{V_{4\text{-}wire}}{I} = R_{RTD} \tag{4.13}$$

From Eqs. (4.10) and (4.13), assuming that the 4-wire measurement is accurate, the error due to 2-wire measurement is given as:

$$R_{measured\text{-}2wire} - R_{measured\text{-}4wire} = R_{wires} \tag{4.14}$$

Example 2 Find out the percentage errors in the 2-wire measurements for the following two cases:

1. Sensor resistance $= 100\ \Omega$ and wire resistance $= 100\ m\Omega$
2. Sensor resistance $= 1\ \Omega$ and wire resistance $= 100\ m\Omega$

Solution
Case 1

$$\% \text{ error contributed by wire resistance} = \frac{100\ m\Omega}{100\ \Omega} = 0.1\ \%$$

Case 2

$$\% \text{ error contributed by wire resistance} = \frac{100\,\text{m}\Omega}{1\,\Omega} = 10\,\%$$

As can be seen in the above example, the error contribution due to 2-wire measurement depends on the sensor resistance to be measured. Hence, in the absence of 4-wire measurement, the 2-wire measurement can still be used by carefully designing the sensor resistance so that the percentage error due to wire resistance is minimized.

4.1.2.5 RTD Calibration

The RTD is calibrated in an isothermal liquid bath chamber (Solbrekken and Chiu 1998) as shown in Figs. 4.9 and 4.10. The objective is to maintain a known and uniform temperature in the package during calibration. A thermally conducting dielectric liquid is used in the bath chamber. A Platinum (Pt) probe is used to measure the temperature of the bath chamber. The package is attached to the test board for the electrical connection (Fig. 4.9). The test board is connected to the data acquisition system. 4-wire method is used to measure the resistance of the RTD. Multiple temperature points of the bath chamber are used for the calibration and a resistance-vs-temperature curve is generated for all the sensors. A typical resistance-vs-temperature curve is shown in Fig. 4.11.

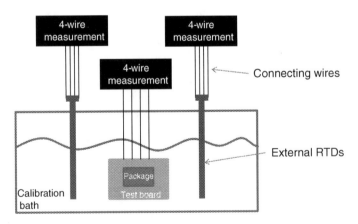

Fig. 4.9 Bath calibration schematic. Package and the test board with wires connected are immersed in the bath. External RTDs are used to measure the temperature of the bath

Fig. 4.10   Bath calibration
chamber

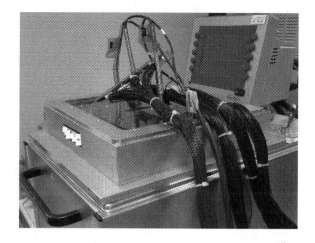

Fig. 4.11   Resistance versus
temperature plot of RTDs
generated from multi-point
bath calibration. Different
slopes correspond to
different samples used in the
calibration

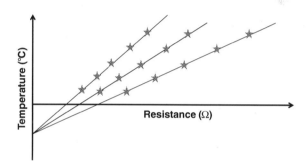

The resistance-versus-temperature relationship shown in Fig. 4.11 is linear. The
linearity depends on the type of the material being used for the RTD and its characteristics with respect to the temperature. The relationship between the resistance
and temperature is given by Eq. (4.15):

$$R = R_0[1 + \alpha(T - T_0)] = R_0(1 + \alpha \Delta T) \qquad (4.15)$$

where,

R is the resistance at temperature T

R_0 is the resistance at reference temperature T_0

α is the temperature coefficient of resistance, also referred as TCR. The TCR is
given by Eq. (4.16):

$$\alpha = \frac{R - R_0}{R_0 \Delta T} = \left[\frac{R - R_0}{R_0}\right]\frac{1}{\Delta T} \qquad (4.16)$$

The first term in Eq. (4.16), $\left[\frac{R-R_0}{R_0}\right]$, represents the percentage or ppm change in the resistance. The unit of α is expressed as ppm/K.

Since multiple data points are required to get the relationship between the RTD resistance and temperature, and create the plot in Fig. 4.11, multi-point calibration is required. The plot in Fig. 4.11 has 5-point calibration.

4.1.2.6 Single Point Calibration

Multi-point calibration takes significant amount of measurement time. A large number of samples are required for the validation of the thermal characteristics of the package. It is not feasible to do multi-point calibration for all the parts or large number of parts. Furthermore, it is not recommended to use the same part for the actual package thermal testing which has undergone the bath calibration and might have been contaminated. A simplified approach called single point calibration technique (Solbrekken and Chiu 1998) is used to solve this problem.

From Eq. (4.16), ΔT can be expressed as given below:

$$\Delta T = \frac{R - R_0}{R_0 \alpha} \tag{4.17}$$

or,

$$T - T_0 = \frac{R - R_0}{R_0 \alpha} \tag{4.18}$$

or,

$$T - T_0 = \left[\frac{1}{R_0 \alpha}\right] R - \left[\frac{1}{\alpha}\right] \tag{4.19}$$

Assuming the reference temperature is $T_0 = 0\,°C$, Eq. (4.19) can be further simplified as shown below:

$$T = \left[\frac{1}{R_0 \alpha}\right] R - \left[\frac{1}{\alpha}\right] \tag{4.20}$$

or,

$$T = mR + c \tag{4.21}$$

where, m is the slope of the curve as shown in Fig. 4.12, and is given as:

$$m = \left[\frac{1}{R_0 \alpha}\right] \tag{4.22}$$

c is the intercept of the curve in Fig. 4.12, and is given as

$$c = -\left[\frac{1}{\alpha}\right] \tag{4.23}$$

From Eq. (4.22), the slope depends on the resistance of the RTD at reference temperature which can vary from one sample to another due to manufacturing variation. However, from Eq. (4.23) it can be seen that the intercept depends on the temperature coefficient of resistance, α, which is the material property of the sensor material and is expected to remain constant from one RTD to another across multiple parts as long as the same material is used in the manufacturing process for the fabrication of RTD. With this assumption, it is possible to use single point to generate the linear calibration curve as shown in Fig. 4.12. The single point can be any known reference temperature. One example could be to use room temperature as the reference temperature. At a room temperature, the resistance of the RTD can be measured using the 4-wire method. This gives the single point data on the temperature vs resistance plot. The intercept, assuming it is constant for all other parts, can be obtained from the multi-point calibration done separately in a different set of sub-samples. Connecting the two points (single point at room temperature and intercept) gives the temperature versus resistance curve. This curve can be generated for all the parts measured where the intercept remains constant and the room temperature resistance varies from part to part due to manufacturing variation.

Since this technique requires just the measurement of the RTD resistance at the room temperature, it can be done in situ during the package thermal testing resulting in an efficient and quick testing (Solbrekken and Chiu 1998) which is feasible for the high volume setup. Another advantage of the single point calibration is that the part is used in the same way as it works in the normal operating condition and is not required to be immersed in any kind of isothermal bath for the calibration.

For the DTS, the two point calibration is usually done—one at low temperature and another one at a high temperature ($T_{j\text{-}max}$). Multi-point calibration becomes prohibitive on a manufacturing line with millions of units been fabricated. Thus,

Fig. 4.12 Single point calibration curves with known intercept. The slope of the curve varies from sample to sample

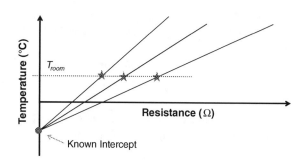

Fig. 4.13 Stack up of a test thermal solution on a microprocessor package during manufacturing

during the test process, the voltage reported by the DTS is assigned to the temperature of the test thermal solution which is used as the true temperature reference. A linear fit is used over the two points that are measured and the information is then fused to the microprocessor. Test thermal solutions that provide accurate DTS calibration are designed such that the contact resistance between the thermal head and the microprocessor is sufficiently low as to meet the requirement of ± 5 °C. A schematic of a test thermal solution on a microprocessor is shown in Fig. 4.13.

A test thermal solution is used to both mechanically load the microprocessor on a test socket as well as to provide cooling of the device during test. To calibrate the temperature sensor of the microprocessor, the temperature of the thermal solution is used as the known temperature reference. To ensure that the temperature sensors meet the required specifications of accuracy, parameters described above are optimized during the manufacturing flow.

Example 3 Given that $R_{room\ temp} = 800\ \Omega$, $\alpha = 4000$ ppm/K, what would be the resistance at 100 °C. Assume room temperature to be 25 °C and the resistance to be linearly dependent on the temperature.

Solution
From Eq. (4.15), R at 100 °C can be estimated as

$$R = R_0[1 + \alpha(T - T_0)]$$

or

$$R = 800 \times \left(1 + (4000 \times 10^{-6}) \times (100 - 25)\right) = 1040\ \Omega$$

Example 4 Use single point calibration to find out die junction temperature, T_j. Given: $R_{room\ temp} = 800\ \Omega$, $\alpha = 4000$ ppm/K, $R_{junction\ temp} = 1000\ \Omega$. Assume room temperature to be 25 °C and the resistance to be linearly dependent on the temperature.

Solution
The T_j can be evaluated with the known slope and intercept as shown in Fig. 4.14. From Eq. (4.21), temperature as a function of the resistance is given as

$$T = mR + c$$

Fig. 4.14 Schematic for T_j evaluation using single point calibration

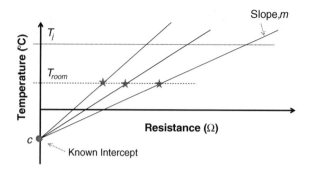

The intercept c can be evaluated using Eq. (4.23) as

$$c = -\left[\frac{1}{\alpha}\right] = -\frac{1}{4000 \times 10^{-6}} = -250\,^\circ C$$

At room temperature

$$T_{room} = mR_{room} + c$$

or,

$$25\,^\circ C = m \times 800\,\Omega - 250\,^\circ C$$

or,

$$m = \frac{25\,^\circ C + 250\,^\circ C}{800\,\Omega} = 0.34375\,^\circ C/\Omega$$

With the known slope m and intercept c, the junction temperature can be calculated as

$$T_j = m \times R_{junction} + c$$

or,

$$T_j = 0.34375\,^\circ\frac{C}{\Omega} \times 1000\,\Omega - 250\,^\circ C = 93.75\,^\circ C$$

4.1.3 Hot Spot ΔT Evaluation

Direct measurement of the hot spot temperature is not feasible as explained in the earlier sections. There can be a large difference in temperatures between the hot spot temperature and the DTS output as shown in Fig. 4.15. Accurate estimation of ΔT between the hot spot and the DTS is another big challenge. It requires theoretical estimation of the hot spot temperature, through modeling and simulations, and generating the temperature distribution across the die for a known microprocessor power and the workload. This serves two purposes.

Fig. 4.15 Die temperature distribution for a given workload. The ΔT exists between the DTS readings and the actual T_j value. ΔT can be computed by estimating the T_j value and measuring the DTS

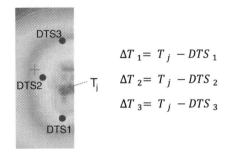

$$\Delta T_1 = T_j - DTS_1$$

$$\Delta T_2 = T_j - DTS_2$$

$$\Delta T_3 = T_j - DTS_3$$

1. It helps in defining the thermal specification (TDP) based on the estimated T_j value.
2. It allows to compute the ΔT between the theoretically estimated hot spot temperature and the measured DTS output.

The modeling and simulations to compute the T_j takes into account the overall thermal management of the microprocessor including the thermal design of the package and the cooling solutions. A detail description of the thermal management methodologies and the computation of product T_j are provided in the subsequent section.

4.2 Microprocessor Thermal Management

The microprocessor junction temperature depends on its thermal management and the type of the cooling system used. A combination of modeling and experimental approach is used to estimate T_j. From Eq. (4.5), for a known P_{total}, ψ_{ja}, and T_{la}, the T_j can be evaluated with the help of modeling. The junction to ambient thermal resistance ψ_{ja} constitutes two elements—(1) package thermal resistance, and (2) system thermal resistance.

Package thermal resistance is an important parameter for the thermal evaluation of the microprocessor. Therefore, it is necessary to understand the details of the thermal resistance and the way it is being defined and used in the semiconductor industries.

4.2.1 Package Thermal Resistance—θ Parameter (θ_{xy})

A typical package assembly will have a stack of components as shown in Fig. 4.16. The parameter θ_{xy} is the thermal resistance between x and y as given in Eq. (4.24), where x and y could be any layer in the stack (Fig. 4.17). The key assumptions are that the package has a uniform power source and all heat goes from x to y.

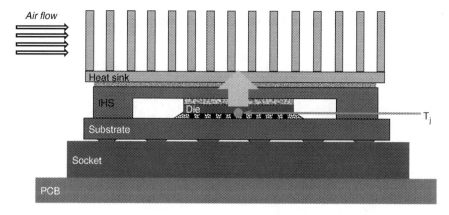

Fig. 4.16 Components in a typical package stack

Fig. 4.17 Nomenclature of components in a package stack where θ_{xy} measurement may be of interest

X and Y could be	
A	Ambient
B	PCB board
C	Case
J	Junction

$$\theta_{xy} = \frac{\Delta T_{xy}}{q} = \frac{L}{kA} \qquad (4.24)$$

where,

ΔT_{xy} is the temperature difference between x and y
q is the input power
L is the length or thickness of the sample
k is the thermal conductivity
A is the cross-section area of the sample
θ_{xy} has the unit of $°C/W$

These assumptions make the definition of θ_{xy} unique in the context of industrial application. θ_{xy} will be compared with the other definitions of package resistances commonly used in the electronics packaging industry in the subsequent sections.

The advantages of using θ_{xy} are: (1) standardized way of measuring thermal resistances, (2) no dependence of system environment, (3) helpful in characterizing the properties of the individual layer of the stack, such as TIM characterization in a standard TIM tester, and (4) useful in generating compact model with equivalent stack properties which in turn is leveraged for system level thermal simulation.

4.2.2 Package Thermal Resistance—R Parameter (R_{xy})

The parameter R_{xy}, also called thermal impedance, is area-independent thermal resistance and is defined as shown in Eq. (4.25).

$$R_{xy} = \frac{\Delta T_{xy}}{q''} = \frac{\Delta T_{xy}}{\frac{q}{A_{source}}} = \frac{\Delta T_{xy}}{q} A_{source} = \theta_{xy} A_{source} = \frac{L}{k} \qquad (4.25)$$

where

A_{source} is the area of the power source
q'' is the heat flux
R_{xy} has the unit of $°C\,cm^2/W$

The assumptions are that the power is uniform over A_{source} and all heat goes from x to y. Note that the R_{xy} is not dependent on the area of the heat source and is only a function of the length/thickness and the thermal conductivity of the layers. R_{xy} is a very popular metric among package thermal analysts and is frequently used for the junction to case thermal resistance (Fig. 4.18) characterization of the package called R_{jc}.

The advantage of using R_{xy} or R_{jc} is that it enables the package characterization and comparison over multiple package form factors with similar stack configuration. This helps package engineers in tracking the thermal performance from one generation to another and from one product segment to another. This also enables the quantification of any improvement in package thermals. However, the disadvantage of R_{xy} is that it cannot be used for non-uniform power source.

4.2.3 Package Thermal Resistance—ψ Parameter (ψ_{xy})

The thermal resistance between x and y with no assumption of all heat going from x to y is given by ψ_{xy} as shown in Eq. (4.26). It is not required to have a uniform power source to compute ψ_{xy}.

$$\psi_{xy} = \frac{\Delta T_{xy}}{q} = \frac{L}{kA} \qquad (4.26)$$

where, ψ_{xy} has the unit of $°C/W$.

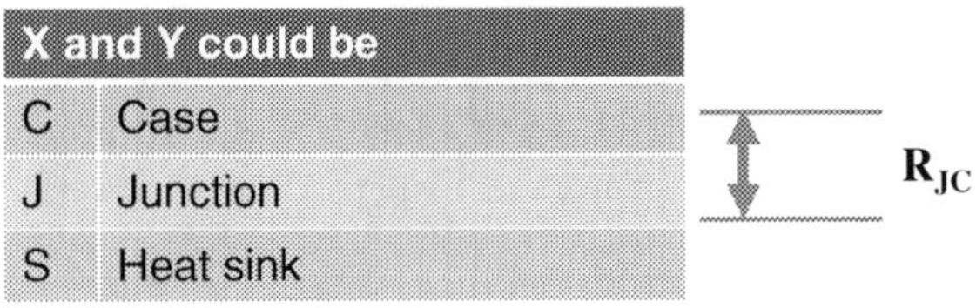

Fig. 4.18 Junction to case thermal resistance R_{jc}

X and Y could be	
C	Case
J	Junction
S	Heat sink

R_{JC}

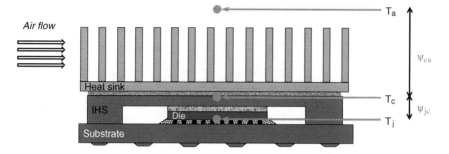

Fig. 4.19 Package stack configuration showing the thermal resistances between junction to case (ψ_{jc}) and case to ambient (ψ_{ca})

The parameter ψ_{xy} is one of the most important parameters for an IC package engineer. It helps in estimating the junction temperature of the silicon die where the power source is generally non-uniform. The advantage of ψ_{xy} lies in the convenience of accurately measuring ΔT_{xy} and the total die power q regardless of what percentage of the total power is dissipated die-up and how much non-uniform the power source is.

For a CPU package where the power source is highly non-uniform and not all the power is dissipated in one direction, ψ_{ja} from junction to ambient is used to estimate the CPU junction temperature. However, for high power packages, it is assumed the most of the power goes die up through the heat sink. Equation (4.5) can be rewritten as

$$T_j = P_{total} \times \psi_{ja} + T_{la} = P_{total} \times (\psi_{jc} + \psi_{ca}) + T_{la}$$

where,

$$\psi_{ja} = \psi_{jc} + \psi_{ca}$$

ψ_{jc} is the junction to case thermal resistance
ψ_{ca} is the case to ambient thermal resistance

The junction to ambient thermal resistance ψ_{ja} constitutes two elements—junction to case thermal resistance ψ_{jc} (package thermal resistance) and case to ambient thermal resistance ψ_{ca} (system thermal resistance). The case temperature, T_c, is defined as the top center temperature of the integrated heat spreader (IHS) as shown in Fig. 4.19.

4.2.4 Thermal Test Vehicle (TTV) Design

A thermal test vehicle (TTV) is used to experimentally characterize the package thermal resistance. A TTV is similar to a microprocessor package and consists of a substrate, silicon die or test chip, TIM, and IHS as shown in Fig. 4.20. The individual components of a TTV are designed to capture the electrical, thermal and mechanical behavior of the microprocessor package. The thermal resistance parameter commonly used to characterize a TTV is R_{jc}. This is because the input power is kept

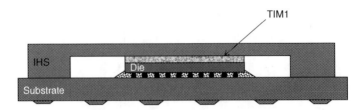

Fig. 4.20 A thermal test vehicle package consisting of substrate, die, TIM1 and the integrated heat spreader (IHS)

uniform, the vertical stack thermal resistance required is independent of the die size, and the TTV is purposely tested in a system where all the heat goes die up. This helps in accurate thermal characterization of the package materials (such as TIM) by eliminating the effects of non-uniform power, die size, and heat split between die up and die bottom. However, in some cases, the TTV can also be designed to simulate the non-uniform power and the heat split that the microprocessor product would experience. In that case, ψ_{jc} is used as the thermal resistance parameter.

Figure 4.21 shows the normalized contribution of the individual components of the package in the thermal resistance. The total package thermal resistance, R_{jc} is primarily due to the thermal resistance of the TIM, R_{TIM} (Prasher 2006). The TTV helps in the understanding and the quantification of the R_{jc} and R_{TIM}.

Substrate: A substrate serves as the electrical and mechanical connection between the mother board and the silicon chip. The TTV substrate dimensions, thicknesses, material of construction, number of dielectric and metal layers, and the metal contents in each layer, and even the process of substrate fabrications are designed similar to that of the final product package substrates.

Silicon test chip: The silicon die is designed with a heater and temperature sensors. Heater is designed to achieve uniform or non-uniform heating depending upon the application to simulate the heating pattern of the microprocessor product. Multiple temperature sensors are required to measure the temperature of the die at different locations. Figure 4.22 shows the different configurations of the heater

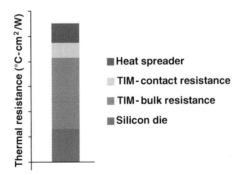

Fig. 4.21 Normalized contributions of thermal resistances from individual components of a typical package. TIM (bulk resistance) is the largest contributor in the stack. Bottom to top shown in the figure—silicon die, TIM (bulk resistance), TIM (contact resistance), and Heat spreader

(a)   **(b)**

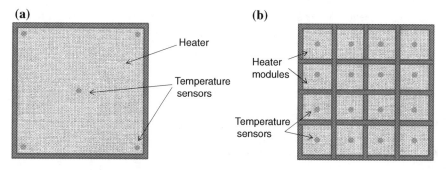

Fig. 4.22 **a** Single uniform heater covering the entire die. The temperature sensors are located based on the TV characterization requirement. **b** Multiple heater modules with dedicated temperature sensors for each module. Each heater module can be either connected with other modules in a series-parallel combination or powered up individually

design and the typical layout of temperature sensors. For uniform heating single heater module is preferred. Multiple heater modules give the flexibility to use the test chip for uniform as well as non-uniform heating. For non-uniform heating, individual modules are powered separately with varying amount of powers to generate the non-uniform heating. To use the multiple modules for uniform heating, the heaters are connected in series and parallel combination thereby keeping the same current through each heater, assuming that the electrical resistance of each heater is the same. However, it is difficult to achieve the same electrical resistance for all the modules due to manufacturing challenges. This results in some error in the final measurement. The objective is to keep this error as low as possible while designing the test chip.

The heating element is made of a long serpentine metallic or any electrically conducting material that is compatible with the CMOS process. Figure 4.23 shows

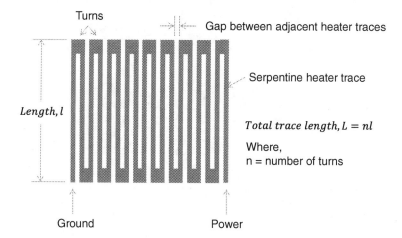

Fig. 4.23 Schematic of a serpentine heater trace normally used for the TV heater design

a serpentine metal trace of a typical heater used in a test chip. The electrical resistance R_{heater} (Ω) of the serpentine metal trace is given by Eq. (4.27)

$$R_{heater} = \frac{\rho L}{A} \qquad (4.27)$$

where,

ρ is the electrical resistivity of the trace material (Ω m).
L is the entire length of the serpentine structure (m) as shown in Fig. 4.23.
A is the cross section area of the trace (m^2).

For a given trace material, R_{heater} can be changed by varying L and A. The value of the designed R_{heater} is normally derived from the power delivery system (Goh et al. 2006). Using the combination of Eqs. (4.28) and (4.29) as detailed below, R_{heater} can be selected.

$$R_{heater} = \frac{P}{I_{max}^2} \qquad (4.28)$$

$$R_{heater} = \frac{V_{max}^2}{P} \qquad (4.29)$$

where,

I_{max} is the maximum current that can be delivered by the power supply.
V_{max} is the maximum voltage that can be delivered by the power supply.
P is the power needed in the test chip heater.

The R_{heater} can be selected to meet the requirements of I_{max}, V_{max} and P. The circuit to power up the heaters is shown in Fig. 4.24. For multiple heater modules, uniform heating can be achieving by using series-parallel combination of the heaters as shown in Fig. 4.24b. For non-uniform heating, each individual heater is powered up separately as shown in Fig. 4.25. This enables creating hot spots in the non-uniform power distribution during the testing. This also allows changing the location of the hot spot and the total power in the heaters depending upon the

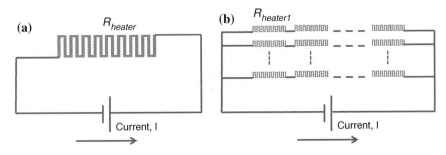

Fig. 4.24 Circuits for realizing uniform heating. Heaters can be powered up through **a** a single heater, or **b** multiple heater modules connected in series-parallel combination

Fig. 4.25 Circuit for realizing non uniform heating. Heater modules can be powered up individually or by connecting in series-parallel combination. The non-uniform heating can be achieved by adjusting the power levels of each heater module

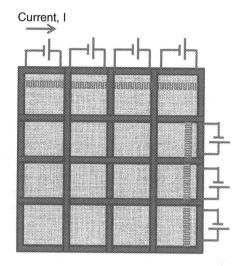

application. However, this can also require significantly high number of package connection pins to be made available in the package design. To obviate that, sometimes a balanced approach is adopted by using some heaters in a series-parallel combination while leaving others to be powered up independently.

The effective resistances of the heater configuration shown in Fig. 4.24b can be estimated by using equivalent resistances of the series and parallel resistors as shown in Eqs. (4.30) and (4.31).

$$R_{series} = R_{heater1} + R_{heater2} + \cdots \qquad (4.30)$$

$$\frac{1}{R_{parallel}} = \frac{1}{R_{heater1}} + \frac{1}{R_{heater2}} + \cdots \qquad (4.31)$$

The actual resistance of the heater is different from the designed heater resistance due to manufacturing variation. For accurate thermal characterization of the package, it is necessary to accurately measure the heater power which in turn is dependent on the accurate measurement of the heater resistance. Similar to the heater design, the temperature sensor design for test vehicles are usually serpentine resistive structure based on the principles of RTD. 4-wire resistance measurement is used for both heater resistance and RTD resistance measurement.

4.2.5 Uniform Power Source and R_{jc} Evaluation

A microelectronics package has multiple heat transfer paths as shown in Fig. 4.26. The dominant heat transfer path will be governed by the type of the products and the end application. In a high power server or high performance computing

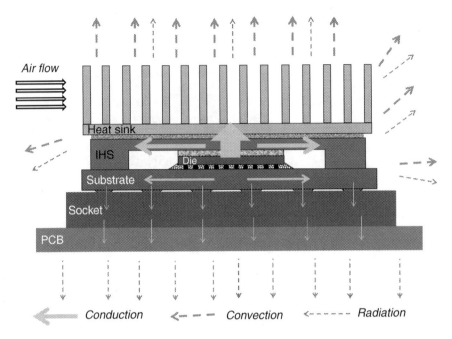

Fig. 4.26 A schematic of heat transfer paths in a typical package

products, the bulk of the heat loss takes place through the heat sink (die-up). In a passively cooled handheld product, a significant portion of the heat loss can be through the substrate and the boards, depending upon the package configuration. The key is to understand the heat flow paths and quantify the heat dissipation in the dominant paths. This can be done by constructing thermal resistance network. A general schematic of thermal resistance network for a typical package is shown in Fig. 4.27.

A package with a uniform power source is required for the TTV. As explained in the previous section, a TTV is used to characterize the thermal performance of the package. It is desired to have negligible or minimum sensitivity towards the input power source which can come from non-uniformity. Hence the first step in the design process of TTV is to achieve uniform power source. The next step is to define a metric which can be used for thermal characterization.

For a CPU package, one of the most important metrics is junction to case thermal resistance R_{jc}. The thermal performance of a package is governed by the magnitude of R_{jc}. There are three elements contributing in R_{jc}—CPU silicon die, thermal interface materials (TIM), and heat spreader. The biggest variable is the TIM which is highly dependent on the process optimization, material selection, and thermo-mechanical effects of the package. R_{jc} essentially captures the TIM performance for an assembled package given that the CPU silicon die and the heat spreader remains approximately the same from package to package and from generation to generation.

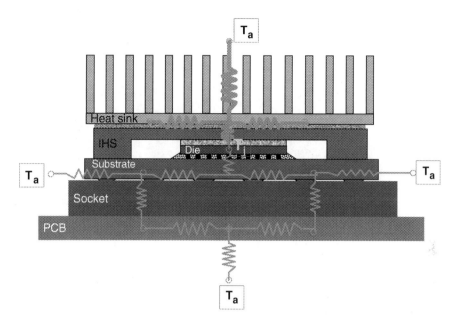

Fig. 4.27 Thermal resistance network for an assembled package

In a high power CPU product almost all heat goes up. To simulate similar effect, a TTV is generally designed to get all the heat going up (Fig. 4.28). This can be achieved by minimizing the die-up thermal resistance. Die-up thermal resistance has two components—junction to case and case to ambient thermal resistances. Junction to case (R_{jc}) is an unknown parameter here, but case to ambient thermal resistance which results from the heat sink and its assembly on the package, can be purposely designed to ensure that the total die-up thermal resistance is minimized and maximum heat flows in the die-up direction. With this setup, R_{jc} can be measured for multiple samples and compared to analyze the package performance.

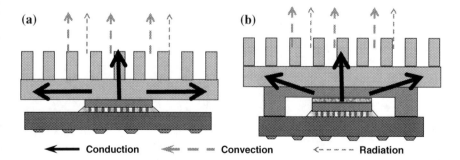

Fig. 4.28 Heat flow path in a typical CPU package for a R_{jc}—**a** package with no integrated heat spreader, **b** package with integrated heat spreader

The Rjc is given as shown in Eq. (4.32)

$$R_{jc} = \frac{T_j - T_c}{P_{uniform}} \cdot A_{die} \tag{4.32}$$

R_{jc} ($°\text{C cm}^2/\text{W}$) is area independent. Using Fourier's law, Eq. (4.32) can be written as

$$R_{jc} = \frac{T_j - T_c}{P_{uniform}} \cdot A_{die} = \frac{l}{k} \tag{4.33}$$

where l is the length or thickness of the stack between junction to case and k is the effective thermal conductivity. The Eq. (4.33) shows that the Rjc depends on the material property. With the known l and k, the R_{jc} can be directly estimated without conducting any measurement. However, given the complexity of the package, it is difficult to accurately estimate the l and k, and hence the measurement of R_{jc} is almost always preferred while characterizing the TTV.

4.2.5.1 Bond Line Thickness

The total stack thickness between the junction and case consists of the thicknesses of die, thermal interface material and the heat spreader. The thickness of the thermal interface material is defined as bond line thickness (BLT) and is shown in Fig. 4.29. The two bounding sides of the TIM interfaces, TIM to IHS and TIM to die, can have voids because of the inability of the TIM to completely wet the surfaces (Prasher 2006). The BLT is not uniform across the die mainly due to package warpage or variation in the flatness of the components. It is this uncertainty that makes the accurate estimation of the BLT quite difficult. There is no single BLT that can accurately capture the package thermal performance.

4.2.5.2 Contact Resistances

The total thermal resistance of the TIM includes the contact resistances at the bounding surfaces (Fig. 4.30) and is given as (Prasher 2006, 2001)

$$R_{TIM\text{-}Effective} = R_{TIM\text{-}bulk} + R_{c1} + R_{c2} \tag{4.34}$$

or,

$$R_{TIM\text{-}Effective} = \frac{BLT}{K_{TIM}} + R_{c1} + R_{c2} \tag{4.35}$$

where,

R_{c1} is the thermal contact resistance at TIM to IHS interface
R_{c2} is the thermal contact resistance at TIM to die interface
K_{TIM} is the bulk thermal conductivity of the TIM

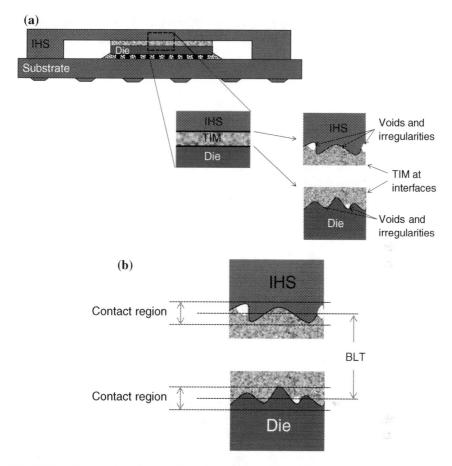

Fig. 4.29 **a** Cross-section of the package showing the voids and irregularities at the TIM interfaces. **b** Contact regions and BLT definition of the TIM

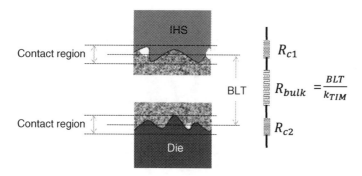

Fig. 4.30 Contact resistances at the two interfaces of the TIM resulting in different total thermal resistance of the TIM compared to the bulk thermal resistance

The thermal contact resistances mainly result from the irregularities and voids at the two interfaces. For more details refer to (Prasher 2006, 2001).

4.2.5.3 T_c Measurement in TTV

The T_c is the temperature at the center of the top of the IHS. It is measured under assembled condition (Fig. 4.31) and a thermocouple is used for the temperature measurement. The T_c is measured when the die is powered on. During testing, the values of T_j or T_c are maintained in the similar range as it would occur in the real product. For an air cooled system, the speed of the cooling fan is controlled to achieve the desired T_j or T_c. Before performing the T_c measurement, a careful process is used to attach the thermocouple at the right location. Figures 4.32, 4.33 and 4.34 shows the schematic of the T_c attach processes.

4.2.5.4 R_{jc} Measurement in TTV

The silicon heater is powered on to generate uniform heating. Thermal solution on top of the package is used to simulate the real operating condition. Under steady state, the T_j and T_c are measured. With the known heater power and the die size (assuming the entire die is covered with the uniform heater), the R_{jc} can be measured in a TTV using Eq. (4.36).

$$R_{jc} = \frac{T_{j\text{-}measured} - T_{c\text{-}measured}}{P_{uniform\text{-}measured}} \cdot A_{die} \qquad (4.36)$$

where,

$T_{j\text{-}measured}$ is the output of the die RTD
$T_{c\text{-}measured}$ is the output of the thermocouple

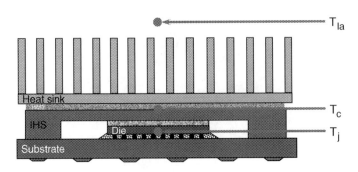

Fig. 4.31 Schematic showing the location of the Tc measurement at the top center of the IHS in the package stack

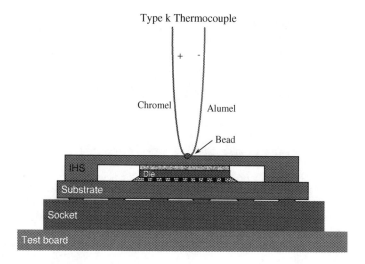

Fig. 4.32 A thermocouple is shown with the bead touching the test specimen where the temperature is measured

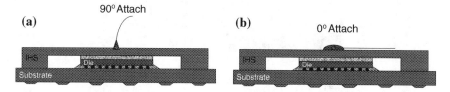

Fig. 4.33 Thermocouple attachment mechanism. **a** The 90° attachment where the thermocouple is attached perpendicular to the test specimen. **b** The 0° attachment where the thermocouple is attached horizontal to the test specimen

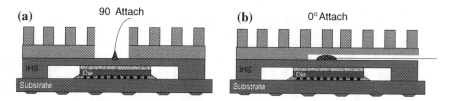

Fig. 4.34 Thermocouple attachment through the heat sink. **a** The 90° attach, and **b** The 0° attach

A quick verification of the measured R_{jc} can be done by theoretically estimating the R_{jc} using 1-D conduction model.

Example 5 Consider the following package configuration with a silicon die attached to a thermal plate (heat spreader) with a thermal interface material (TIM). The die is dissipating power and the heat is lost from the thermal plate surface.

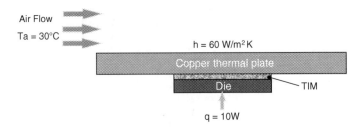

Fig. 4.35 An electronic package configuration with air flow at the *top* of the thermal plate

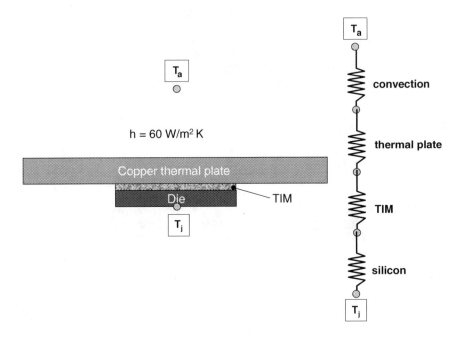

Fig. 4.36 Equivalent thermal resistance circuit of the package described in Fig. 4.35

What is the die temperature (T_j) and the junction to case thermal resistance (R_{jc})? Assume 1-D conduction.

Copper thermal plate:

Thermal conductivity, $k = 400$ W/mK
Size = 50 mm × 50 mm
Thickness = 5 mm

Silicon die:

Thermal conductivity, $k = 120$ W/mK
Size = 15 mm × 15 mm
Thickness = 0.5 mm

TIM:

Thermal conductivity, k = 1 W/mK
BLT = 0.05 mm

Solution

Construct the equivalent thermal circuit as shown in Fig. 4.36 and evaluate the thermal resistances for each segment of the package using Eq. (4.24).

$$\theta_{conv} = \frac{1}{hA_s} = \frac{1}{60 \times 50 \times 50 \times 10^{-6}} = 6.66 \,^{\circ}C/W$$

$$\theta_{plate} = \frac{L}{KA} = \frac{0.005}{400 \times 50 \times 50 \times 10^{-6}} = 0.005 \,^{\circ}C/W$$

$$\theta_{TIM} = \frac{L}{KA} = \frac{0.05 \times 10^{-3}}{1 \times 15 \times 15 \times 10^{-6}} = 0.22 \,^{\circ}C/W$$

$$\theta_{silicon} = \frac{L}{KA} = \frac{0.5 \times 10^{-3}}{120 \times 15 \times 15 \times 10^{-6}} = 0.02 \,^{\circ}C/W$$

The total thermal resistance between junction to case can be given as

$$\theta_{ja} = \theta_{silicon} + \theta_{TIM} + \theta_{plate} + \theta_{convection} = 6.9 \,^{\circ}C/W$$

Using Eq. (4.24), θ_{ja} is related to T_j as given below.

$$\theta_{ja} = \frac{T_j - T_a}{P} = 6.9 \,^{\circ}C/W$$

From the above equation, the unknown T_j can be evaluated as

$$T_j = (6.9 \,^{\circ}C/W)(P) + T_a = (6.9 \,^{\circ}C/W)(10\,W) + 30\,^{\circ}C = 99\,^{\circ}C$$

From Eq. (4.25), R_{jc} can be computed as

$$R_{jc} = \frac{T_j - T_c}{P_{uniform}} \cdot A_{die} = \frac{l}{k}$$

$$R_{jc} = \frac{l_{silicon}}{k_{silicon}} + \frac{l_{TIM}}{k_{TIM}} + \frac{l_{plate}}{k_{plate}}$$

The contact resistances between the TIM and the thermal plate as well as the TIM and the die interfaces have been assumed to be negligible.

$$R_{jc} = \frac{0.5 \times 10^{-3}}{120} + \frac{0.05 \times 10^{-3}}{1} + \frac{5 \times 10^{-3}}{400}$$

or,

$$R_{jc} = 6.67 \times 10^{-5} \,^{\circ}C\,m^2/W = 0.667 \,^{\circ}C\,cm^2/W$$

4.2.6 Non-uniform Power Source and ψ_{jc} Evaluation

Microprocessor packages under normal operation have a non-uniform power sources and ψ_{jc} is used to define the thermal specifications. The ψ_{jc} can be either measured or computed. To power on a microprocessor for conducting the measurement, it has to be assembled with the motherboard along with the cooling solution. The setup resembles a perfectly working computer system. For the ψ_{jc} measurement, a given workload is run on the system and the measured ψ_{jc} is corresponding to that particular workload. After the system is powered on, the T_j and T_c are measured under steady state condition. The ψ_{jc} is given as shown below.

$$\psi_{jc} = \frac{T_{j\text{-}measured} - T_{c\text{-}measured}}{P_{non\text{-}uniform\text{-}measured}} \tag{4.37}$$

where,

$T_{j\text{-}measured}$	is the output of the die DTS
$T_{c\text{-}measured}$	is the output of the thermocouple
$P_{non\text{-}uniform\text{-}measured}$	is the total power supplied to the microprocessor.

When a TDP workload is run the $P_{non\text{-}uniform\text{-}measured}$ is equivalent to TDP. Under this case, the $T_{j\text{-}measured}$ can approach $T_{j\text{-}max}$ assuming that the rest of the boundary conditions approach the designed specified conditions.

The ψ_{jc} can also be computed by using a 3D model. The thermal specifications are finalized based on the computed ψ_{jc} long before the product is ready for the testing. The inputs for the 3D model require a non-uniform powermap and the thermal properties of the package. A non-uniform powermap is theoretically estimated and validated in a separate experiment. The package thermal properties are extracted from the TTV data by measuring R_{jc}.

4.2.6.1 Density Factor (DF)

The impact of the non-uniformity can be quantified by a metric called the density factor (DF) (Torresola et al. 2005). The DF is a ratio of thermal resistance to thermal impedance for a given package configuration and is given as

$$DF = \frac{Thermal\ Resistance\ (\psi_{jc})}{Thermal\ Impedance\ (R_{jc})} \tag{4.38}$$

where,

$$\psi_{jc} = \frac{T_j - T_c}{P_{non\text{-}uniform}} \tag{4.39}$$

$$R_{jc} = \frac{T_j - T_c}{P_{uniform}} . A_{die}$$

(4.40)

The unit of DF is cm^{-2}. The DF can be computed by using 3D package modeling where all the boundary conditions for the two cases (ψ_{jc} and R_{jc}) are kept the same except for the power inputs. In case of the ψ_{jc} model, the power input is non-uniform whereas for R_{jc} it is uniform.

4.2.6.2 Thermal Impact of Non-uniform Power

A sample comparison between uniform power and non-uniform power thermal resistances is shown in Fig. 4.37. The thermal resistance between junction to ambient is higher for a non-uniform case, which in turn reduces the cooling capability of a particular heat sink. In other words, the overall TDP is reduced. If the cooling capability is 100 W under uniform power air cooled system thermal solution, then with the non-uniform power air cooled system, the cooling capability decreases to 80 W. The cooling capability can be increased by using an improved system cooling solution such as liquid cooling, instead of air cooling (Fig. 4.37). The percentage of package thermal resistance (ψ_{jc} or R_{jc}) compared to the total thermal resistance (ψ_{ja}) increases by approximately 2X from uniform case to non-uniform case and by 3X from uniform air cooled case to non-uniform liquid cooled case. It is quite evident that the package thermal resistance dominates the total stack resistance. For this reason a lot of research focus has been given to improve the package thermal resistance especially the TIM1 resistance. Advances in TIM1 materials, characterization and testing have

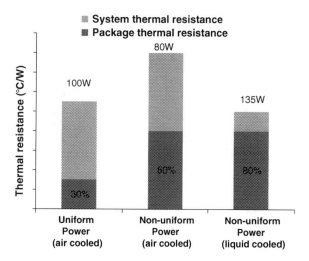

Fig. 4.37 Sample comparison of cooling capability between uniform power and non-uniform power thermal resistances. Contribution of package thermal resistance increases from 30 to 50 % from uniform power to non-uniform power (air cooled) thereby increasing the overall thermal resistance and thus impacting the cooling capability from 100 to 80 W

been widely reported in the literature (Prasher 2006; Samson et al. 2005; Russell and Chiu 1998; Chiu et al. 1997, 2000, 2001, 2002; Solbrekken et al. 2000).

Example 6 Find out the cooling capability (TDP) of a microprocessor under different non-uniformities (DF) of the input powers. Given: $R_{jc} = 0.25\,^{\circ}\text{C}\,\text{cm}^2/\text{W}$, $\psi_{ca} = 0.5\,^{\circ}\text{C}/\text{W}$, $T_{j\text{-}max} = 100\,^{\circ}\text{C}$, $T_{la} = 30\,^{\circ}\text{C}$. Evaluate it for multiple density factors: DF $= 1, 2, 3,$ and 4.

Solution
Known parameters:

$$T_{j\text{-}max} = 100\,^{\circ}\text{C}$$
$$\psi_{ca} = 0.5\,^{\circ}\text{C}/\text{W}$$
$$R_{jc} = 0.25\,^{\circ}\text{C}\,\text{cm}^2/\text{W}$$
$$T_{la} = 30\,^{\circ}\text{C}$$

From Eq. (4.3), TDP can be computed as

$$TDP = \frac{T_{j\text{-}max} - T_{la}}{\psi_{ja}}$$

or,

$$TDP = \frac{T_{j\text{-}max} - T_{la}}{\psi_{jc} + \psi_{ca}}$$

Using Eq. (4.38), the TDP can be written as

$$TDP = \frac{T_{j\text{-}max} - T_{la}}{DF * R_{jc} + \psi_{ca}}$$

or,

$$TDP = \frac{100 - 30}{DF * 0.25 + 0.50} = \frac{70}{DF * 0.25 + 0.50}$$

For the DF of 1, 2, 3, and 4, the TDPs are shown below.

DF (cm^{-2})	TDP (W)
1	93
2	70
3	56
4	47

As can be seen from the above results, the cooling capability of a microprocessor decreases as the non-uniformity of the input power increases. The DF of 1 is equivalent to the uniform power.

4.3 Sensitivity Analysis and Thermal Impacts

As discussed before in Eqs. (4.1) and (4.2), the TDP is directly related with the frequency of the microprocessor and can be written as shown below.

$$TDP = P_{dynamic} + P_{leakage} = C_{dyn} V_{cc\_max}^2 f + P_{leakage} \tag{4.41}$$

or,

$$f = \frac{TDP - P_{leakage}}{C_{dyn} V_{cc\_max}^2} \tag{4.42}$$

Frequency is one of the key performance metrics of the microprocessor. The objective is to achieve higher frequency of the product to gain the performance. At the same, it is also preferred to have minimum TDP to reduce power dissipation while achieving the frequency target. For most of the high performance computing microprocessors, the maximum allowed power dissipation is limited by the cooling system of the product. That means for a given TDP, a product generally has a fixed maximum operating frequency and hence performance. An increase in TDP results in the enhancement of the performance. However, an error or guard bands in the T_j measurement and estimation can result in the conservative estimate of TDP causing negative performance impact which could have been utilized otherwise.

Any known bias and variation in the temperature measurement and the package characterization is accounted for as a guard band in the T_j. The T_j is conservatively specified in the thermal specification which could have otherwise translated into the higher specified performance in the product specification. To quantify the thermal performance benefits achievable by improving the accuracy and precision, a sensitivity analysis is required. From Eq. (4.3), TDP can be expressed in terms of the metrics that are either measured or modeled as given in equations below. The errors associated with these metrics will directly impact the TDP.

$$TDP = P_{dynamic} + P_{leakage} = \frac{T_{j-max} - T_{la}}{\psi_{ja}} \tag{4.43}$$

The ψ_{ja} is comprised of ψ_{jc} and ψ_{ca}. The ψ_{jc} and ψ_{ca} are separately analyzed. This is because ψ_{jc} is estimated by the package designer whereas the ψ_{ca} is estimated by the system designer. The package can be used by multiple system designers with different types of cooling solutions. The technical specification for the package designer requires it to meet the T_j limit whereas the technical specification of the system designer requires it to meet T_c limit specified by the package designer for the given sets of boundary conditions. The above equation can be written in terms of ψ_{jc} and ψ_{ca} as given below.

$$TDP = \frac{T_{j-max} - T_{la}}{\psi_{jc} + \psi_{ca}} \tag{4.44}$$

Fig. 4.38 Sensitivity plot
of TDP with respect to
the density factor (DF) for
different values of R_{jc}

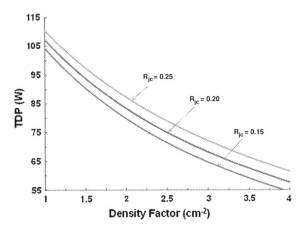

The ψ_{jc} is theoretically estimated by using 3D model and given by the following equation.

$$\psi_{jc\text{-}estimated} = \frac{T_{j\text{-}estimated} - T_{c\text{-}estimated}}{P_{non\text{-}uniform\text{-}estimated}} \qquad (4.45)$$

Using Eq. (4.38), ψ_{jc} can be written in terms of R_{jc}.

$$\psi_{jc} = DF \cdot R_{jc} \qquad (4.46)$$

The R_{jc} is measured on the TTV and is given as

$$R_{jc} = \frac{T_{j\text{-}measured} - T_{c\text{-}measured}}{P_{uniform\text{-}measured}} \qquad (4.47)$$

By combining Eqs. (4.46) and (4.47), ψ_{jc} can be written as

$$\psi_{jc} = DF \cdot \frac{T_{j\text{-}measured} - T_{c\text{-}measured}}{P_{uniform\text{-}measured}} \qquad (4.48)$$

Similarly, the thermal resistance for case to ambient can be written as

$$\psi_{ca\text{-}non\text{-}uniform} = CF + \psi_{ca\text{-}uniform} \qquad (4.49)$$

where, CF is the correction factor and ψ_{ca} is given as

$$\psi_{ca} = \frac{T_{c\text{-}measured} - T_{la\text{-}measured}}{P_{uniform\text{-}measured}} \qquad (4.50)$$

Therefore,

$$TDP = \frac{T_{j\text{-}max} - T_{la}}{\psi_{jc} + \psi_{ca}} = \frac{T_{j\text{-}max} - T_{la}}{DF \cdot R_{jc} + CF + \psi_{ca\text{-}uniform}} \qquad (4.51)$$

The measured and the modeled metrics in Eq. (4.51) are DF and for the sensitivity analysis the $T_{j\text{-max}}$ and T_{la} in Eq. (4.51) can be assumed to be fixed.

$$TDP = \frac{T_{j\text{-max}} - T_{la}}{DF \cdot \frac{T_{j\text{-measured}} - T_{c\text{-measured}}}{P_{uniform\text{-measured}}} + CF + \psi_{ca\text{-uniform}}} \tag{4.52}$$

Assuming typical values for $T_{j\text{-max}}$ (100 °C), T_{la} (25 °C), ψ_{ca}, and CF, a sensitivity plot between TDP and DF can be generated as shown in Fig. 4.38. An error of 20 percent in R_{jc}, which can happen due to variation in the measurement of key metrics like $T_{j\text{-measured}}$, $T_{c\text{-measured}}$, or $P_{uniform\text{-measured}}$, can generate a TDP error of close to 10 percent (Fig. 4.38).

In conclusion, there are multiple factors that are responsible for the thermal characterization of microprocessor and Tj computation. The silicon die temperature sensor plays a crucial role in the entire process where a detailed methodology is followed. The process can become more elaborate as the microprocessor package becomes complex. The next chapter briefly describes some of the complicated package configurations.

References

Borkar, S., "Design challenges of technology scaling", IEEE Micro, July – August (1999).

Borkar, S., Chien, A., "The future of microprocessors," Communication of the ACM, vol. 54, No. 5, May (2011).

Chiu C., Solbrekken G., and Chung Y., "Thermal modeling of grease-type interface material in PPGA application," in Proc. 13th Annu. IEEE Semiconductor Thermal Measurement and Management Symp. (SEMI-THERM), 1997, pp. 57–63 (1997).

Chiu C., Solbrekken G., and Young T., "Thermal modeling and experimental validation of thermal interface performance between non-flat surfaces," in Proc. 7th Intersociety Conf. Thermal and Thermo-Mechanical Phenomena in Electronic Systems (ITHERM 2000), vol. 1, pp. 52–62 (2000).

Chiu C., Maveety J., and Tran Q., "Characterization of solder interfaces using laser flash metrology," Microelectron. Reliab. vol. 42, no. 1, pp. 93–100, Jan. (2002).

Chiu C., Chandran B., Mello M., and Kelley K., "An accelerated reliability test method to predict thermal grease pump-out in flip-chip applications," in Proc. 51st Electronic Components and Technology Conf., 2001, pp. 91–97 (2001).

Goh, T.J., Chiu, C.P., Seetharamu, K.N., Quadir, G.A., and Zainal, Z.A., "Test chip and substrate design for flip chip microelectronic package thermal measurements", Microelectronics International, Vol. 23, Number 2, 2006, 3 – 10 (2006)

Moore, G., "Cramming more components onto integrated circuits," Electronics, vol. 38, pp. 114–117, Apr. 19 (1965).

Prasher, R., "Surface chemistry and characteristic based model for the thermal contact resistance of fluidic interstitial thermal interface materials,". J. Heat transf., vol. 123, pp. 969–975 (2001).

Prasher, R., "Thermal interface materials: Historical perspective, status, and future directions", Proceedings of the IEEE, vol. 94, No. 8, Aug (2006).

Russell A. and Chiu C., "A testing apparatus for thermal interface materials," in Proc. 1998 Int. Symp. Microelectronics, 1998, pp. 1036–1041 (1998).

Samson, E. C., Machiroutu, S. V., Chang, J. Y., Santos, I., Hermerding, J., Dani, A., "Prasher, R., Song, D. W., "Interface material selection and a thermal management technique in second-generation platforms built on Intel Centrino Mobile Technology," Intel Technology Journal, Vol. 9, Issue 1 (2005).

Solbrekken, G.L., Chiu, C.P., "Single point calibration method for die level temperature sensors", IEEE Intersociety Conference on Thermal Phenomenon (1998).

Solbrekken G., Chiu C., Byers B., and Reichebbacher D., "The development of a tool to predict package level thermal interface material performance," in Proc. 7th Intersociety Conf. Thermal and Thermomechanical Phenomena in Electronic Systems (ITHERM 2000), pp. 48–54 (2000).

Torresola J., Chiu C., Chrysler G., Grannes D., Mahajan R., and Prasher R., "Density factor approach to representing impact of die power maps on thermal management," IEEE Trans. Adv. Packag., vol. 28, no. 4, pp. 659–664, Nov. (2005).

Yuffe, M., Knoll, E., Mehalel, M., Shor, J., Kurts, T., "A fully integrated multi-CPU, GPU and memory controller 32nm processor", IEEE International Solid-State Circuits Conference, February 22 (2011).

Chapter 5
Microelectronics Thermal Sensing: Future Trends

Chandra Mohan Jha, Leila Choobineh and Ankur Jain

Microelectronics architecture and packages are becoming increasingly complex with a greater number of components, including 3D stacks, getting integrated into a single package, called multi-chip packages (MCPs). MCPs are being used very widely in the microelectronics industry and enable increasing bandwidth demands (Chong 2012; Polka et al. 2007). The input/output (I/O) bandwidth demands are quite aggressive which result in faster signal data rates, higher signal count, or higher signal density and crosstalk through signal interconnects, which typically dominate high speed signal performance between silicon devices. An MCP places the devices closer in a single package with a shorter interconnect length between the devices and improves the interconnect impedance mismatch compared to a conventional solution of using multiple packages on a board. However, MCPs come with their own thermal, thermo-mechanical, and reliability challenges which in turn affect product performance. Thermal characterization and temperature sensing in a MCP plays an important role in establishing the product performance. This chapter describes the design and engineering details of MCPs with a focus on 3D packages, and presents the challenges with the characterization and the temperature sensing of the die.

C.M. Jha (✉)
Intel Corporation, Santa Clara, USA
e-mail: cmjha75@gmail.com; chandra.mohan.m.jha@intel.com

L. Choobineh · A. Jain
University of Texas at Arlington, Arlington, USA
e-mail: choobineh.leila@mavs.uta.edu

A. Jain
e-mail: jaina@uta.edu

© Springer Science+Business Media New York 2015
C.M. Jha (ed.), *Thermal Sensors*, DOI 10.1007/978-1-4939-2581-0_5

5.1 Multi-chip Packages

MCPs, as shown in Fig. 5.1, are typically complex packages where heterogeneous silicon devices (microprocessor logic, memory, etc.) are integrated.

Microprocessor frequency performance can be improved by increasing its thermal design power (Mahajan et al. 2006). The thermal design power (TDP), as described in the previous chapter, is the maximum sustained microprocessor power corresponding to a set of pre-defined realistic applications, that can be cooled to meet the thermal and reliability constraints of the product. The TDP of the product can be increased by using improved cooling (i.e. reduced thermal resistance) solutions. The cooling capability of a microprocessor depends on the combined effects of the package thermal solution (package substrate + die + thermal interface material (TIM1) + heat spreader) and the system cooling solution (TIM2 + heat sink). TIM1 is the thermal interface between the die and the heat spreader and TIM2 is the thermal interface between the heat spreader and the heat sink. Typical system cooling technologies have been widely covered in the literature (Sauciuc et al. 2005; Chu et al. 2004; Shaukatullah 1999). For a product with non-uniform power maps and local regions of higher power, package thermal resistance can dominate the overall stack thermal resistance. The influence of the package thermal resistance increases in MCP configurations. MCP configurations are typically classified as 2D, 2.5D or 3D (Polka et al. 2007). The higher influence of the package thermal resistance in MCP configuration is primarily due to (1) multiple thermal interface layers in the die stack, and (2) larger variation in the thickness of the interface bond line thickness (BLT) between the top die and the heat spreader compared to a single chip package (SCP). The combined effects of localized non-uniform power and higher package thermal resistance result in hot spots that are no longer limited to single die. The hot spots can be generated anywhere in the package either in-plane (2D or 2.5D configuration) or out-of-plane (2.5D or 3D configuration) depending upon the application. This adds to the challenge of (1) accurately estimating the junction temperature, T_j, and (2) accurately measuring the die temperature closest to the hot spot.

For accurate estimation of T_j, all thermal interfaces between die to die need to be characterized. Figure 5.2 shows a schematic of a single 3D stack MCP with

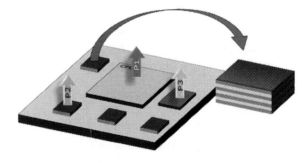

Fig. 5.1 A schematic of an MCP package where heterogeneous devices are co-packaged in a combination of 3D and 2D configurations with different powers shown as *P*1, *P*2, and *P*3

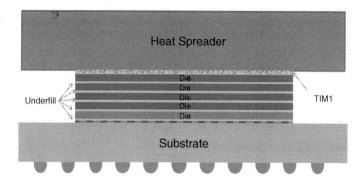

Fig. 5.2 A schematic of a single 3D stack with underfill material between die and TIM1 material between the top die and the heat spreader

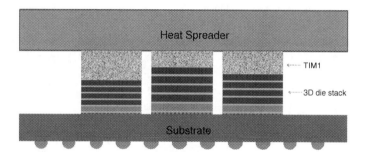

Fig. 5.3 A schematic of a multiple 3D stacks placed side by side

multiple die stacked on top of each other. The stacking provides significant electrical and volumetric benefits, however it increases the overall package thermal resistance due to multiple thermal interface layers. It is assumed that there is an underfill (UF) material between die1 and die2, die2 and die3, and so on. The thermal interface layer between the top die and the heat spreader is called a typical thermal interface material TIM1. Generally UF and TIM1 have different material characteristics, thermal conductivity, and contact resistances, with thermal conductivity of typical commercial UFs lower than that of TIM1.

Figure 5.3 shows another schematic of a multi-chip package where multiple 3D stacks are placed side by side. Variations in the overall die stack thicknesses result in a variation of the TIM bond line thickness (BLT). A thicker BLT increases the package thermal resistance. For MCPs, there can be a variation of the overall heights of the individual die or package stacks on the substrate. Package warpage also contributes to the overall challenge of efficient thermal management. In a package, warpage is driven by the CTE mismatch between the die(s) and the substrate and is also a function of the ratio of total die area to the package substrate area.

The structural coupling of the multiple components attached on the same substrate affect the thermo-mechanical behavior of each component. The dynamic warpage

(warpage at higher temperatures) and thermo-mechanical stresses during the operating lifetime of the component can be higher for multi-component packages.

Given the complexity of these packages with the requirement of several layers of UF and multiple thermal interfaces, the package thermal resistances in the MCPs can be several times higher than that of a conventional single-chip package. As described before, the cooling capabilities in these packages are mainly dictated by the package thermal resistances and to a lesser extent by the system thermal resistances. This places further importance on the accuracy of temperature sensing and the package characterization.

5.2 3D Packages

Much research has been carried out in the past few decades on three-dimensional integrated circuits (3D IC). The basic concept behind 3D IC technology is the realization of multiple parallel planes of active transistors instead of the traditional paradigm of all transistors being on the same plane. While monolithic 3D integration is certainly possible, its application has been restricted due to several processing challenges related to growth of epitaxial Silicon on top of metal features. The preferred approach is to manufacture differentiated die separately, and then carry out integration using die-to-die, die-to-wafer or wafer-to-wafer bonding. Through-Silicon vias (TSVs) and metal micropads are key components of 3D IC technology. 3D ICs have been shown to result in significant improvement in signal transmission and interconnection power compared to traditional microelectronic systems (Banerjee et al. 2001; Patti 2006; Loh et al. 2007). The evolution of 3D IC technology as well as individual components are now important components of ITRS projections (source: http://www.itrs.net/).

The long-term goal of this research field is to realize multi-die, logic-dominant 3D IC stacks. Heat dissipation in such a stack is clearly an important technological challenge. A direct effect of vertical stacking is an increase in the power density, without an improvement in the die footprint from which heat could be removed. Die thinning, an important technology necessary for fine pitch TSVs, also results in additional thermal management difficulties, since a thinner Silicon die has reduced heat spreading capability. Compared to a traditional microelectronic system, physical access to intermediate strata in a 3D IC stack becomes a challenge. Traditionally, one side of Silicon has been used for electrical interconnection, and the other side for cooling. In a 3D IC, however, cooling of most of the intermediate die is a considerable challenge. Finally, the thermal characteristics of components unique to 3D IC technology, such as TSVs and metal micropads continue to be addressed. Since TSVs are made of a metal such as Cu or W, for which the thermal conductivity is greater than that of Silicon, in principle, a TSV may be useful as a thermal conduction pathway, particularly for localized heat generation spots. This capability and its coupled nature with the electrical role of TSVs, as well as stress generation due to a TSV have been investigated. Novel methods such as microfluidic cooling, including the provision of microchannels to access each stratum, solid-state cooling using thermoelectric coolers, etc., are being investigated (Dang et al. 2010; Matthew et al. 2013; Brunschwiler et al. 2009).

While the integrated thermal-electrical design has always been important for microelectronic devices, this is a particularly critical challenge for 3D ICs. Thermal and electrical optimization is more closely coupled than ever before, and the technological penalty of carrying out only electrically-driven system design is greater than ever before. It has been shown that while thermal-only or electrical-only optimizations result in significant electrical or thermal penalty respectively, a coupled co-optimization results in designs that while not optimal from either perspectives, are reasonably close to both optima and are able to combine the best of both the worlds. A significant amount of research is being carried out to understand thermal concerns in 3D ICs (Dang et al. 2010; Matthew et al. 2013; Brunschwiler et al. 2009; Venkatadri et al. 2011; Choobineh and Jain 2013, 2012; Jain et al. 2011; Alam et al. 2010; Cong et al. 2007; Goplen 2006; Kim et al. 2010), and to quantify thermal effects on electrical performance of 3D ICs. The thermal effect of 3D IC process technologies such as wafer thinning, TSV formation, etc. has been investigated in detail (Jain et al. 2011; Alam et al. 2010; Cong et al. 2007; Goplen 2006).

From the perspective of temperature sensing, a 3D IC offers unique challenges. Effective temperature sensing on multiple spots in each die is critical for enabling unique thermal management strategies such as thermoelectric cooling and microfluidic cooling. In addition, off-chip temperature sensing, for example in microfluidic lines, is also important. Temperature measurements on-chip and off-chip need to be integrated and analyzed in order to make resource allocation decisions, such as the relative amount of coolant flowrate to direct to each stratum, etc. Accessibility of temperature sensors interspersed among multiple die, multiplexing of these sensor outputs, the effect of signal transmission through TSVs and metal micropads, etc. are important practical considerations for temperature sensing in 3D ICs. In addition, the modeling of temperature distribution in a 3D IC is important, as this will help interpret thermal measurement data and inform thermal-electrical decision making. Thermal modeling of a 3D IC is considerably more challenging compared to a traditional IC. Since heat generation is now distributed in the vertical direction as well, it is important to model and predict the temperature distribution in each stratum of the 3D IC. Novel package architectures related to 3D IC technology, such as 2.5D interposer packages offer interesting thermal characteristics (Choobineh et al. 2013). The modeling of thermal transport in a 2.5D interposer is also a relevant future research direction for accurate thermal measurements and operation of microelectronic systems.

5.3 Modeling and Simulation Techniques

This section summarizes thermal modeling and simulation for microelectronic systems. The relative pros and cons of various categories of modeling techniques are discussed.

5.3.1 Modeling Methods

The fundamental principle behind the modeling and simulation of temperature fields in any physical system is the conservation of energy. This basic principle is represented by a variety of formulations depending of the system under consideration, and the approach adopted for analysis. For a rectangular coordinate system, the fundamental governing energy equation is given by

$$k \left[\frac{\partial^2 T}{\partial x^2} + \frac{\partial^2 T}{\partial y^2} + \frac{\partial^2 T}{\partial z^2} \right] + \frac{Q}{k} = \rho C_p \frac{\partial T}{\partial t} \tag{5.1}$$

where T is the spatially and time varying temperature distribution, Q is the heat generation rate, and k, ρ, C_p are thermal conductivity, density and heat capacity, respectively. Equation (5.1) already assumes that k is isotropic and temperature-independent, both of which are reasonable assumptions for Silicon close to room temperature. When solving for the steady-state temperature distribution, the transient term on the right-hand side is dropped.

The governing energy equation is associated with appropriate boundary conditions and initial conditions. An adiabatic boundary condition is often assumed for the side-faces of a die, due to the small surface area. Other boundary conditions may represent heat dissipation by the active layer, which in general may be space-dependent, and cooling due to the heat spreader or heat sink, which is often represented by a convective heat transfer coefficient.

The fundamental energy conservation represented by Eq. (5.1) may be solved using a variety of analytical and numerical tools to result in the temperature distribution. Analytical methods include the method of separation of variables, Green's functions, Laplace transforms, particularly for transient problems, etc. A number of reduced-order methods have been investigated (Huang et al. 2004; Joshi 2012) that sacrifice some accuracy for greater computation speed. One particular approach of interest discretizes the physical domain into smaller elements, and writes energy equations involving the temperatures at specific nodes. These equations, similar to nature of Ohm's law that governs flow of electric current, are solved in order to produce the temperature distribution. This approach often utilizes algorithms for solving RC circuits that are well developed due to their application in traditional analysis of electrical circuits (Huang et al. 2004). These methods have been utilized in several recently developed analysis tools such as HotSpot and its variants (Huang et al. 2004).

Finite-element simulation is another class of methods used for determination of the temperature distribution in microelectronic systems. These popular methods are the underlying basis of several commercially available thermal analysis software tools. The relative ease of use forms an important advantage over other methods.

5.3.2 3D Package Modeling

Modeling of temperature distribution in 3D ICs has been carried out using each of the modeling techniques outlined above.

Researchers from the University of Virginia have developed tools capable of temperature computation in a multi-die 3D IC (Huang et al. 2004). HotSpot 3D, which is a variant of their more fundamental temperature computation tool called HotSpot is based on discretization of the physical domain of interest into smaller elements, followed by solving a set of discretized energy conservation equations based on each element (Huang et al. 2004; Joshi 2012).

Finite-element simulation of temperature distribution in 3D ICs has been carried out using numerical analysis tools. Due to the user-friendly nature of these tools, these simulations are relatively straightforward and easy to run. However, like in any other software package, it is important to understand the nature of the underlying computations behind simulation results.

A third direction in which temperature computation of 3D ICs have been recently reported (Choobineh and Jain 2012, 2013) concerns the analytical solution of the fundamental governing energy equations. Two particular analytical methods—iterative and non-iterative—have been presented. In the iterative approach for an N-die stack, the heat flux into each die is guessed, which results in analytical expressions for the temperature distribution in each die. This is in turn used to predict the heat flux into each die. The guessed heat flux distribution is corrected using a weighted average of the previous and current values. The process is then repeated, and in this manner, improvement in the guessed heat flux distribution is obtained over multiple iterations. Figure 5.4 shows the temperature plot obtained for a representative problem over a number of iterations. In order to obtain transient temperature distributions, Laplace transform of the governing transient energy equation is carried out, which results in elimination of the transient term. The resulting equations are solved in the Laplace domain using

Fig. 5.4 Convergence of temperature distribution using the iterative method

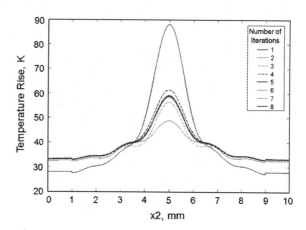

Fig. 5.5 Comparison of temperature computed from the iterative approach with finite-element simulation results

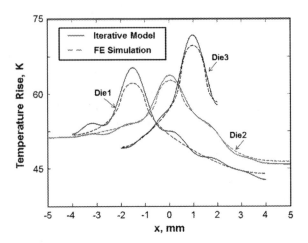

the iterative approach outlined above. Conversion from Laplace domain to time domain is carried out using a technique based on de Hoog's algorithm. Figure 5.5 compares the temperature distribution obtained from this approach to a finite-element simulation result. In both cases, a three-die 3D IC with unequally sized die is considered. A single hotspot is assumed to exist on each die. There is excellent agreement between the iterative model and finite-element simulation results.

Temperature fields determined from an iterative analytical approach agree well with the predictions from finite-element simulation. Typically the accuracy increases by increasing the number of iterations and, in case of transient computations, by increasing the number of terms considered by de Hoog's algorithm for transformation to time domain. It has been found that only 7–8 iterations are sufficient for engineering accuracy.

A second approach for analytical solution of temperature distribution in 3D ICs is based on a non-iterative approach that derives an explicit expression for the temperature distribution. In this case, the temperature distribution is written as a Fourier series expansion using eigenfunctions in the in-plane x and y directions, whereas the temperature variation in the vertical z direction is accounted for by z-dependent coefficients. The coefficients are determined by inserting the assumed form of the solution into the boundary conditions in the vertical direction. If perfect thermal contact between adjacent strata is assumed, the coefficients in the z-direction can be determined exactly. In case a certain thermal contact resistance exists between strata, a set of linear algebraic equations in the unknown coefficients are obtained. A number of standard approaches are available to solve such equations. Similar to the iterative method, results from this approach agree well with predictions based on finite-element simulations, as shown in Fig. 5.6.

The analytical methods outlined above have been utilized to compute temperature distribution in 3D ICs with equally sized die, and also with unequally sized die. Temperature distributions for non-uniform heat generation on each die have been obtained. For example, using the non-iterative approach outlined above, the

Fig. 5.6 Comparison of temperature computed from the non-iterative approach with finite-element simulation results

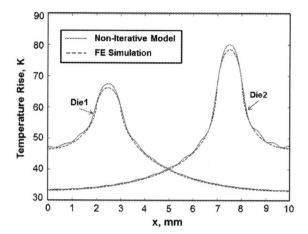

temperature distribution for a 10-die stack with complicated, space-dependent heat generation can be computed rapidly within several tens of seconds.

In addition, the iterative approach has also been modified in order to analyze a 2.5D interposer structure, in which multiple small-sized die are bonded on a single large carried substrate, often referred to as the interposer. In this case, the heat flux from each die into the interposer is guessed, resulting in the solution of the temperature distribution, which is in turn used to improve upon the guessed heat flux distribution. Multiple iterations result in convergence of the temperature distribution.

5.4 Conclusions

This chapter briefly summarized the future directions and challenges in thermal engineering of microelectronics, and the accompanying needs and challenges in thermal sensing. Thermal modeling and sensing must keep pace with advances in die-level microelectronic architectures as well as packaging. In particular, 3D integration is identified as a promising technology for which innovative thermal modeling and sensing techniques will be needed in the future.

References

Ang Boon Chong, "Multi Chip Packaging (MCP) or Not MCP?", Proc of the International MultiConference of Engineers and Computer Scientists (IMECS), vol. II, Mar. 2012

Alam, S.M., Jones, R.E., Pozder, S., Chatterjee, R., Jain, A., 'Interstratum connection design considerations for cost-effective 3-D system integration', *IEEE Trans VLSI Systems*, **18**(3), pp. 450-460, 2010. (DOI: 10.1109/TVLSI.2008.2011910).

Banerjee, K., Souri, S.J., and Saraswat, K. C., '3-D ICs: A novel chip design for improving deep submicron interconnect performance and systems-on-chip integration', *Proc. IEEE*, **89**(5), pp. 602-633, 2001.

Brunschwiler T., Michel B., Rothuizen H., Kloter U., Wunderle B., Oppermann H. and Reichl H., "Interlayer Cooling Potential in Vertically Integrated Packages," *Microsystem Technol.*, **15**(1), pp. 57-74, 2009.

Choobineh, L., Jain, A., 'Analytical solution for steady-state and transient temperature field in vertically integrated three-dimensional integrated circuits (3D ICs)', *IEEE Trans Components, Packaging & Manufacturing Technologies,* **2**(12), pp. 2031-2039, 2012. (DOI: 10.1109/TCPMT.2012.2213820).

Choobineh, L., Jain, A., 'Determination of temperature distribution in three-dimensional integrated circuits (3D ICs) with unequally-sized die', *Applied Thermal Engineering*, **56**, pp. 176-184, 2013. (DOI: 10.1016/j.applthermaleng.2013.03.006).

Choobineh, L., Agonafer, D., Jain, A., 'Analytical modeling of temperature distribution in interposer-based microelectronic systems', Proc. ASME/IEEE InterPACK 2013, San Francisco.

Chu, R.C., Simons, R.E., Ellsworth, M.J., Schmidt, R. R., Cozzolino, V., 2004, "Review of cooling technologies for computer products," Device and Materials Reliability, IEEE Transactions, Vol. 4, Issue 4, pp. 568 – 585.

Cong, J., Luo, G., Wei, J. and Zhang, Y.: 'Thermal-aware 3D IC placement via transformation', *Proc. Asia & South Pacific Design Automation Conf.*, pp. 780-785, 2007.

Dang, B., Bakir, M.S. ; Sekar, D.C. ; King, C.R. ; Meindl, J.D., 'Integrated Microfluidic Cooling and Interconnects for 2D and 3D Chips,' *IEEE Trans. Adv. Packaging*, **33**, 2010, pp. 79-87.

Goplen, B. and Sapatnekar, S. S., 'Placement of thermal vias in 3D ICs using various thermal objectives,' *IEEE Trans. Comput.-Aided Design of Integr. Circuits Syst.*, **26**(4), pp. 692–709, 2006.

Huang W., Stan M. R., Skadron K., Sankaranarayanan K., Ghosh S., and Velusamy S., "Compact Thermal Modeling for Temperature-Aware Design." In Proceedings of the 41st Design Automation Conference, June 2004.

Joshi, Y., 'Reduced Order Thermal Models of Multiscale Microsystems,' *J. Heat Transfer*, **134**(3), 2012, 031008.

Jain, A., Alam, S.M., Pozder, S., Jones, R.E., 'Thermal–electrical co-optimisation of floorplanning of three-dimensional integrated circuits under manufacturing and physical design constraints', *IET Computers & Digital Techniques* (Special Issue on Three-dimensional Integrated Circuits Design), **5**(3), pp. 169-178, 2011. (DOI: 10.1049/iet-cdt.2009.0107).

Kim Y.J., Joshi Y.K., Federov AG., Lee Y.J., Lim S.K., "Thermal Characterization of Interlayer Microfluidic Cooling of Three-Dimensional Integrated Circuits with Nonuniform Heat Flux," *ASME J. Heat Transfer*, **132**, 041009-1, 2010.

Loh, G., Xie, Y., and Black, B.: 'Processor design in three-dimensional die-stacking technologies', *IEEE Micro*, **27**(3), pp. 31-48, 2007.

Mahajan, R., Chiu, C. P., & Chrysler, G. (2006). "Cooling a microprocessor chip." Proceedings of the IEEE, 94(8), 1476-1486.

Matthew, R., Manickaraj, K., Sullivan, O., Mukhopadhyay, S. and Kumar, S., "Hotspot Cooling in Stacked Chips using Thermoelectric Coolers," IEEE Transactions on Components and Packaging Technologies, 3 (5), 759-767, 2013.

Patti, R.: 'Three-dimensional integrated circuits and the future of system-on-chip designs', *Proc. IEEE*, **94**(6), pp. 1214-1224, 2006.

Polka, L. A., Kalyanam, H., Hu, G., Krishnamoorthy S., "Package Technology to address the memory bandwidth challenge for Tera-scale computing", Intel Technology Journal, vol. 11, issue 03, August 2007.

Shaukatullah, H., 1999, "Bibliography of liquid cooled heat sinks for thermal enhancement of electronic packages," Semiconductor Thermal Measurement and Management Symposium.

Sauciuc, I., Prasher, R., Chang, J. Y., Erturk, H., Chrysler, G., Chiu, C. P., Mahajan, R., "Thermal performance and key challenges for future CPU cooling technologies." Proceedings of IPACK 2005, ASME InterPACK '05, July 17-22, Sanfrancisco, California, USA.

Venkatadri, V., Sammakia, B., Srihari, K., Santos, D., 'A review of recent advances in thermal management in three dimensional chip stacks in electronic systems', *ASME J. Electronic Packaging*, **133**, pp. 041011-1, 2011.

Chapter 6
Thermal Sensors for Energy Converter Applications

S.P. Duttagupta, P. Ramesh, S. Roy, R.A. Shukla, S.G. Kulkarni and G.J. Phatak

Portable electronic devices of today and the future will require increasingly sophisticated, efficient, micro and nano-scale power sources. Such a power source may harvest energy even from human saliva. These devices are often temperature sensitive and may require an integrated thermal management system (I-TMS) for optimized operation. The proposed I-TMS will comprise an array of heaters and thermal sensors. The goal is to generate uniform heating throughout the system and monitoring of the same.

S.P. Duttagupta (✉) · P. Ramesh · S. Roy · R.A. Shukla
Indian Institute of Technology Bombay, Mumbai 400076, India
e-mail: sdgupta@ee.iitb.ac.in

P. Ramesh
e-mail: ramp@ee.iitb.ac.in

S. Roy
e-mail: sandipta.r@iitb.ac.in

R.A. Shukla
e-mail: nandan.shukla@gmail.com

S.G. Kulkarni · G.J. Phatak
Centre for Materials for Electronics Technology (C-MET), Panchavati,
off Pashan road, Pune 411008, India
e-mail: kulkarnishri29@gmail.com

G.J. Phatak
e-mail: gjp@cmet.gov.in

© Springer Science+Business Media New York 2015
C.M. Jha (ed.), *Thermal Sensors*, DOI 10.1007/978-1-4939-2581-0_6

6.1 Thermal Management for Micro Energy Converters

The technology trend towards miniaturization as well as advanced packaging solutions have resulted in significant advances in field-deployed sensors and electronic communication. The size and weight of the portable appliances are decreasing and at the same time the accuracy and resolution has improved. A critical issue is the durability and reliability of the localized power source, which involves energy storage and energy conversion.

Accordingly, a significant research and development effort has been directed towards development of micro-scale energy storage and converter (ES&C) systems. These ES&C systems need to be highly scalable, to micron dimensions if possible. Such systems are also required to be versatile—provide a large amount of power over a short period of time (burst-mode application)—in addition to the more conventional requirement of continuous low power output. While new battery technologies are being explored, there is also considerable interest in development of micro-fuel cells as high efficiency energy converters (Kulkarni 2014; Mitra 2014; Patil et al. 2014; Ramesh 2014; Dimble 2012; and Kulkarni et al. 2012).

Recently there has been significant interest in application of wireless sensors network (WSN) for Integrated Vehicular Health Management (IVHM) for automobiles, aircrafts, and unmanned vehicles (air-borne, water-borne). The IVHM requirements include monitoring of temperature, deformation, and vibrations in a distributed manner. Due to the extreme environment application, all electronics and power source must be rated for operation in the temperature range of 200–450 K. A second requirement is the monitoring of exhaust gases which requires the sensors and power sources to be operated in the temperature range of 900–1200 K. The precision measurement of temperature is critical for the safe and reliable functioning of the sensors, power sources, (micro) computers, communication and controls systems which are all part of the IVHM portfolio.

The localized power sources noted above can be a battery or even a piezo-ceramic based energy harvesting unit. Micro fuel cells are being explored as a promising energy conversion device (Zhao 2009). Fuel cells (current source) are able to meet load transient demands more effectively than a battery. They are also not affected by the charge-discharge cycle effects and are thus potentially more durable. Finally, fuel cells have the unique advantage of being able to operate over a wide range of temperatures as required for IVHM type applications. This chapter describes the design of micro-heaters and micro temperature sensors as part of an **Integrated Thermal Management System (I-TMS)** for micro fuel cells.

Fuel cells are electrochemical devices that convert the chemical energy of a reactant directly into electrical energy. Among the types of fuel cells that are being explored, **Micro Proton Exchange Membrane Fuel Cells (M-PEMFC)**, having certain advantages such as, faster start-up, higher power density, compactness, light weight and low operating temperature. Such M-PEM fuel cells can be used as a distributed, micro-scale power source (Lin et al. 2008; Madou et al. 2006). The operational temperature for PEM fuel cells is typically 350 K. For ultra-high

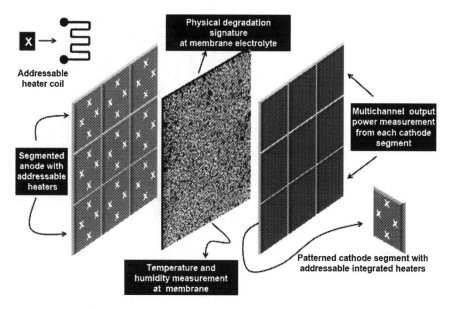

Fig. 6.1 Micro fuel cell schematic with segmented anodes and cathodes, and integrated, addressable micro-heaters and micro thermal and humidity sensors (reproduced from Ramesh 2014)

temperatures (1000 K) **Micro Solid Oxide Fuel Cells (M-SOFC)** are being explored (Huang 2009) which will require integrated heaters and temperature sensors to perform at that temperature level.

The various components of an advanced micro fuel cell are (i) Gas inlet and outlet flow structures, (ii) Gas diffusion layers, and (iii) Membrane Electrode Assembly (MEA) comprising of the two electrodes (cathode, anode) and the intermediate ion conducting electrolyte. A segmented M-PEMFC with distributed heaters has been developed (Ramesh 2014) (Fig. 6.1) which resulted in uniform heating and which helped in mitigating the effects of local hotspots (temperature transients). Advanced M-PEMFC designs can incorporate micro heaters and micro thermal sensors as part of an integrated thermal management system (Ramesh 2014). Such an I-TMS design can be extended for high temperature M-SOFC devices as well. The incorporation of I-TMS is required to create fuel cells with high conversion efficiency, extended operational life and reliable output for application as an energy source for field deployed sensor networks (Ramesh 2014).

Some of the details to realize 3-D integration of various micro fuel cell components including micro-heaters and micro thermal sensors are illustrated in the schematic in Fig. 6.1 (Ramesh 2014). The objective is to design an intelligent system which allows for *extraction of maximum power even under non-optimal conditions.* Another important design objective is to enhance and *extend operational lifetime by protecting against transients* generated at the input/output as well as from within the fuel cell, all of which contributes to cell degradation.

Fig. 6.2 Micro proton
exchange membrane (PEM)
fuel cell (reproduced from
Dimble 2012)

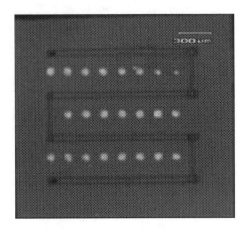

Fig. 6.3 Micro PEM fuel
cell output (current–voltage,
current–power) characteristics
(reproduced from
Dimble 2012)

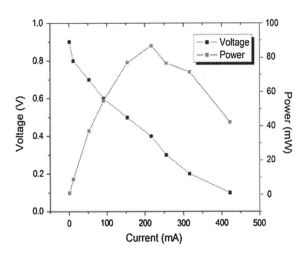

Improvement in cell efficiency (short-term) and reliability (long-term) are especially critical for remote or in vivo power source applications.

A first generation 3-D packaged micro fuel cell (M-PEMFC) is shown
(Fig. 6.2) as well as the current versus voltage (I–V) and current versus power
(I–P) characteristics (Fig. 6.3). The active cell dimension was approximately
1 cm², hence the maximum recorded power density is about 90 mW/cm².

The various parameters of interest are (i) start-up time, (ii) power density,
(iii) compactness, (iv) packaged mass, (v) operating temperature, and (vi) operational
lifetime. Multi-sensor nodes have been shown to be useful for wide-area monitoring and control for operation of re-configurable solar photo-voltaic systems (Patnaik
et al. 2014; Patnaik et al. 2012; Patnaik et al. 2011), smart power-grids (Duan 2012),
and for chemical plume monitoring in remote and possibly hostile environments
(Roy et al. 2014).

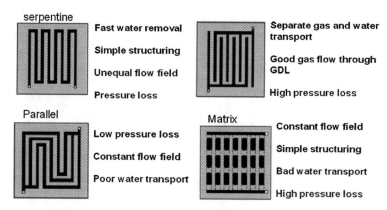

Fig. 6.4 Variable flow field patterns for bipolar plates used for fuel distribution in a PEM fuel cell (reproduced from Ramesh et al. 2013)

The design of M-PEMFC was implemented based on a 3-D model using a COMSOL Multi-physics simulation platform (Patil et al. 2013; Dimble et al. 2011; Ramesh et al. 2012; Ramesh et al. 2011). The membrane (electrolyte) is commercially available form of Poly-Tetra-Fluoro-Ethylene (PTFE).

Significant effort was involved in the design and optimization of electrodes (cathode, anode) and catalyst (Platinum nano-composite) and to form an integrated Membrane Electrode Assembly (MEA) that is the heart of a PEM fuel cell (Dimble 2012). Adjacent to the MEA are the Gas Diffusion Layers (GDL) on the cathode as well as anode side. The fuel (for example, hydrogen) and oxygen are transported to the GDL via flow-fields created using micro-channels etched in Stainless Steel or Graphite. Multiple micro-channel designs are shown in Fig. 6.4.

Next, the fuel cells components are assembled following a 3-D layer by layer process using Low Temperature Co-fired Ceramic (LTCC) Technology as illustrated in Figs. 6.5 and 6.6. A cavity based fabrication and packaging process can provide high quality external electrical contacts without damaging the membrane electrode assembly (Ramesh 2014).

The micro-fuel cell packaging is comprised of multiple layers (Fig. 6.6). The L1 and L2 are cavity layers. A gas diffusion layer (GDL) and membrane electrode assembly (MEA) are placed in these layers. Two layers are used in order to match the thickness of cavity with the thickness of the GDL and MEA. L3 and L4 are used for flow field and current vias. Two layers are used in order to obtain the desired channel height and increase mechanical strength of the device. Layer L5 contains two apertures for fuel inlet and outlet. The L6 is the heater layer and can also accommodate a thermal sensor such as a thermistor. The L7 layer is used to increase the mechanical strength of the package.

A technique for direct collection of current from the Gas Diffusion Layer (GDL) has been developed (Ramesh 2014). The current collectors are arranged in a distributed fashion (Mitra et al. 2013; Patil et al. 2013; Mitra et al. 2012;

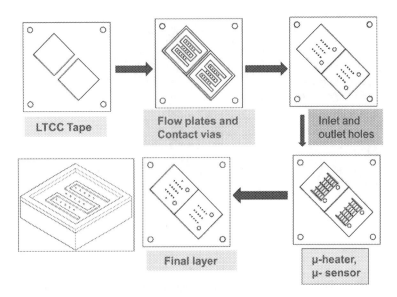

Fig. 6.5 Micro fuel cell components assembly using LTCC technology (reproduced from Ramesh 2014)

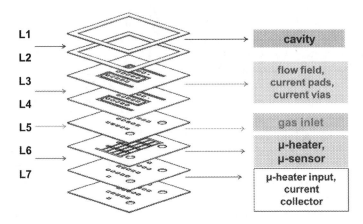

Fig. 6.6 Schematic of layer by layer assembly of M-PEMFC components (reproduced from Ramesh 2014)

Ramesh et al. 2011; Lin et al. 2008). It has been observed that a number of small area cells connected in parallel is superior to a single large area cell (Ramesh 2014).

The extended operation (a few weeks) of a M-PEMFC in the field, where cathode is fed by ambient air (in place of high quality oxygen), has been studied using an in-house developed test setup (Ramesh 2014). The effect of air-borne contaminants on long-term fuel cell performance has been reported (Patil et al. 2014; Ramesh et al. 2013). Fuel starvation conditions have been carefully analyzed (Patil et al. 2013; Ramesh et al. 2011; Modav et al. 2006) and mitigation strategies

(Mitra et al. 2012) have been proposed. The presence of local hotspots (Ramesh et al. 2012) in the membrane electrode assembly can lead to irreversible degradation (Mitra et al. 2012). A segmented micro fuel cell with distributed micro-heaters has been fabricated which helps mitigate degradation (Ramesh 2014). Presently, efforts are ongoing to integrate micro thermal sensors in the 3-D package.

6.2 Thermal Sensors for Micro Fuel Cells

The micro thermal sensor utility for micro fuel cells is to monitor operational temperatures as well as the uniformity of heating of the membrane electrode assembly (MEA) of a micro fuel cell. The fuel cell is highly sensitive to temperature variations due to the impact on the half-cell reactions at the electrodes (cathode and anode), as well as the ion flux through the membrane. Heating uniformity is a factor because most of the MEA components (for example, the PTFE membrane) are poor conductors of heat. The nature of distributed heating is expected to minimize thermal transients that cause fluctuations in the ion flux (and hence output power). Any hot-spot formation also has a deleterious impact on the operational life-time of the fuel cell.

The temperature bands are specific to a particular fuel cell but may vary considerably depending on the fuel cell technology under consideration. For example, proton exchange membrane (PEM) fuel cells typically operate at temperatures less than 400 K. On the other hand, the maximum operating temperatures for a solid oxide fuel cell (SOFC) may reach 1300 K. As a result, no single thermal sensor technology will be suitable for all temperature bands. Also subject to consideration is the ability to integrate sensors using the Low Temperature Co-fired Ceramic (LTCC) packaging process.

In the next sections, three different micro thermal sensor technologies for I-TMS for fuel cell application are described. The first is a ceramic based Negative Temperature Coefficient (C-NTC) sensor which is suitable for monitoring M-PEMFC devices (low temperature operation). The second is an Avalanche Photo Diode (APD) based sensor, which relies on temperature dependent break-down voltage as the operating principle. Finally, for M-SOFC devices (high temperature operation), a metal-semiconductor-metal based photo-detector has been developed which can be used to remotely estimate the temperature by recording the Infra-Red (IR) emission from the micro-heater element incorporated in the fuel cell.

6.3 Ceramic Micro-Heater, Micro-Sensor for M-PEMFC

The operation of micro-PEM fuel cell requires integrated micro heater and temperature sensor operating in the band of 350–400 K. Ceramic heaters as well as sensors are quite appropriate for this temperature range. Ceramic heaters can also be utilized for the high temperature band (900–1300 K) operation of micro SOFC. The advantage here is that a highly compact, all-ceramic cell design can be used, if all component elements (fuel cell, heater, and sensor) are compatible with the LTCC process flow.

6.3.1 Integrated Micro Heater for PEMFC and SOFC

For micro-PEMFC, micro heater is packaged using Low Temperature Co-fired Ceramics (LTCC) technology, whereas for SOFC all components are ceramic based. LTCC offers many potential advantages for fabrication, assembly and packaging such as miniaturization, portability, faster response time, and low energy consumption for such devices (Vasudev et al. 2013). LTCC demonstrates excellent chemical inactivity, high temperature stability, bio-compatibility and mechanical strength and possibility of complex 3-D structure construction. There are a few reports on the micro heater for the gas sensor applications in LTCC for Liquefied Petroleum Gas (LPG), smoke or other reducing gases (Jain et al. 2007; Pisarkiewicz et al. 2003; Golonka et al. 2006). The temperature range achieved is a maximum of 800 K, the output power varies from 250 mW to 3 W, and the heaters are either buried or on-surface. For fuel cell application the heater typically remains buried, while the sensors may operate in contact (thermistors, APDs) or in non-contact mode (photo-detectors). The focus is to integrate multiple heater and sensor elements into one 3-D package.

The micro heater element fabrication has been reported using different techniques. These include screen printing (Pisarkiewicz et al. 2003), post-deposition laser patterning (Nowak et al. 2009), and laser based filling of patterns (MolDovan et al. 2007). The patterns are either serpentine or straight lines (Jain et al. 2007; MolDovan et al. 2007). The heater materials are conducting ceramic (RuO_2), Platinum (Jain et al. 2007) or alloy of Gold, Platinum, and Palladium (Au/Pt/Pd) (MolDovan et al. 2007). The specific heater design and fabrication process flow is determined by the end use of the micro heaters or the processes available for R&D and manufacturing.

There are studies which have focused on the integrated heater and temperature sensor arrangement (Gongora-Rubio et al. 2001). Several different micro heater designs and fabrication processes are available in LTCC. Suspended screen printed micro heaters, metallic micro heaters and buried resistor heaters have been fabricated using DuPont CF011 and platinum paste. A modified layout schematic (Fig. 6.7) for metal and ceramic based micro-heaters for fuel cell application has been implemented (Kita et al. 2000).

Platinum is used in LTCC structure to screen print high temperature heaters. LTCC packages with these heaters can reach temperature up to 900 K with low power consumption of 10 W. These heaters can be easily screen printed and integrated with LTCC package.

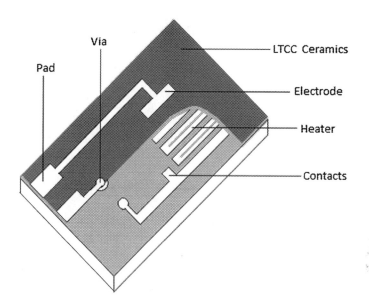

Fig. 6.7 Layout schematic for metal and ceramic based micro-heaters following design proposed by Kita et al. (2000)

6.3.2 Integrated Micro Temperature Sensor for PEMFC

The three most common types of micro-scale, contact temperature sensors in use are: (a) Thermocouples, (b) Resistance temperature detectors (RTDs), and (c) Thermistors.

Ceramic based **thermocouples** have a very small foot-print and are non-intrusive, hence they can be installed in devices where the design constraints do not allow room for accommodation of any additional structure. For example, in case of a gas turbine application, the thickness of the sensor is less than the gas phase boundary layer thickness. Micro-scale, thin-film thermocouples that are directly deposited on the blades and vanes can be very effective in measuring the engine temperature. Thin film thermocouples can provide greater accuracy in the measurement of surface temperatures of such applications due to low thermal inertia, hence less thermal vibrations.

The thermocouple sensitivity is dependent on the electrical properties of its individual elements. Ceramic thermistors that are made from Indium Tin Oxide (ITO) can be fabricated with different thermos-elements which can be used for high temperature applications.

Thermistors are essentially resistors with a high temperature coefficient of resistance (TCR):

$$TCR = \frac{dR_T}{R_T dT}. \tag{6.1}$$

The thermistor resistance R_T changes exponentially with temperature T and can be described by the relationship:

$$R_T = R_0 e^{\frac{B}{T}}. \tag{6.2}$$

The B constant (beta factor or coefficient of thermal sensitivity) is a thermistor parameter which reflects the change in resistance with temperature.

The value of B constant can be determined on the basis of the following formula:

$$B = \frac{T_1 T_2}{T_2 - T_1} \ln \frac{R_1}{R_2}, \tag{6.3}$$

where, R_1 and R_2 are resistances at temperatures T_1 and T_2 respectively.

Thick-film thermistors are prevalent because of low cost and high TCR coefficient, which means high thermal sensitivity. However, they have lower long term stability. High TCR enables manufacture of smaller footprint and higher sensitive components. Due to their low cost, simple construction, high sensitivity and high signal to noise ratio, thermistors are utilized for precision temperature measurement and control. Most common types of temperature sensors are based on NTC thermistor and LTCC resistors.

The most widely used are **ceramic based negative temperature coefficient (C-NTC)** thermistors that comprise of semiconducting ceramics with either spinel or perovskite structure. At higher temperatures, materials with perovskite structure are preferred, instead of conventional ones based on spinel due to unstable composition and structure, and significant aging (degradation in electrical property with thermal cycle).

NTC and other thick-film thermistors fired on the surface or buried inside multilayer LTCC substrates have been subject of research. Typically, slip cast multi-layers of $La_{0.7}Sr_{0.3}Zr_{0.5}Co_{0.2}^{2+}Co_{0.3}^{3+}O_3$ (LSZC) and $La_{0.8}Sr_{0.2}Ti_{0.5}Co_{0.3}^{2+}Co_{0.2}^{3+}O_3$ (LSTC) ceramics with the perovskite structure co-sintered with platinum contacts are used.

In the temperature range from 200 to 700 K, the TCR values vary from $(-)12.6$ to $(-)2$ %/K and from $(-)11.0$ to $(-)1.1$ %/K for LSTC and LSZC thermistors, respectively. The values of B constant ranges from 6300 to 8000 and 4200 to 5100 in this temperature range for LSTC and LSZC thermistors, respectively.

The simulated thermal variation of resistance of a ceramic based negative thermal coefficient (C-NTC) thermistor is shown in Fig. 6.8. The simulations are following the experimental results reported by Dziedzicet al. We observe the decrease in resistance (expressed as a resistance ratio: R/R_{300}), the reference temperature being 300 K) with increase in operating temperature.

The most important C-NTC thermistor parameters are: constant B, tolerance of the constant B and long-term stability. High value of constant B ensures good sensor sensitivity. Repeatability of the sensor parameters depends on the constant B

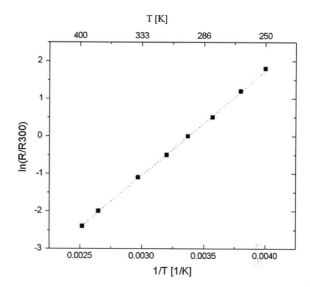

Fig. 6.8 Simulated thermal variation of resistance of a C-NTC thermistor following experimental results reported by Dziedzic et al. (2005)

variability coefficient. In general C-NTC thermistors exhibit weak stability. Most of the commercially available thermistor compositions are matched with alumina substrates. A difference in TCE (Thermal Coefficient of Expansion) between LTCC and alumina substrates may lead to deformation. Moreover, the physicochemical interaction between the tape and the paste materials affects thermistor properties. Surface thermistors reportedly exhibit worse long-term stability than buried ones.

The relative change in resistances are seen even with low ageing (200 h, 370 K). However, changes in thermistor constant B is much lower. Therefore these material systems are suitable to be utilized for temperature sensors in LTCC circuits and ceramic MEMS.

Low resistivity LTCC resistor paste has been reported to provide the best chemical and mechanical compatibility with LTCC devices and have a higher overall stability compared to other two types (Kulkarni 2014).

The stability of embedded resistors is relatively better than that of surface ones. Certain proprietary buried resistors are quite stable, with fractional resistance changes in ±0.3 % range and TCR less than 20 ppm/K.

6.4 IR Micro-Temperature Sensor for M-SOFC

For SOFC technology it will be advisable to use ceramic micro-heaters for typical operational temperatures of 900–1100 K. For this temperature band a different sensing principle is required. Metal-semiconductor-metal (MSM) photo-detectors for non-contact, yet proximate, estimation of heater temperature can be used.

Table 6.1 High temperature
thermocouples

Material (type)	Temperature range (K)
Chromel–constantan (E)	220–1000
Iron–constantan (J)	230–1000
Chromel–alumel (K)	80–1630
Ni/Mo–Ni/Co (M)	100–1680
Pt (70)–Rh (30) (B)	270–2000
Pt (87)–Rh (13) (R)	270–1800
Pt (90)–Rh (10) (S)	270–1830
W (95)–Re (5) (C)	Extreme temp

A metal-semiconductor junction based electronic device (Schottky diodes) has a number of applications ranging from a switch to an infra-red (IR) detector (Rogalski 2003; Schwarz et al. 1996). While a single metal-semiconductor junction is simpler from a fabrication standpoint, a metal-semiconductor-metal (MSM) structure is also in use (Roy et al. 2014). Silicon (band gap 1.12 eV at 300 K) is the traditional choice for semiconductor, there are also references to Germanium (small band gap, 0.67 eV), Gallium Nitride (wide band gap, 3.3 eV), Silicon Carbide (wide band gap, 3.3 eV) have also been reported (Saddow et al. 2004). Specifically, germanium-on-silicon based metal–semiconductor–metal photodetectors, with responsivity of 0.85 A/W at 1.55 μm and for 2 V reverse bias have been demonstrated (Okyay et al. 2006). The photodiode exhibited reverse dark currents of 100 mA/cm^2 and external quantum efficiency up to 68 %. The wide band-gap semiconductor based Schottky diodes have application in detecting Ultra-Violet (UV) and X-ray radiation as well as high-energy alpha particles and thermal neutrons (Das et al. 2012).

The cut-off wavelength (upper limit) of the detection band in the infra-red (as an example) is governed by the difference in metal semiconductor work-function(s) which is known as the barrier height of Schottky diode.

IR photo-detectors can be used to estimate temperatures of micro solid oxide fuel cells operating at 900–1200 K. At present, the only commercially available contact detectors are expensive, high-temperature (for example, Type S) thermocouples (see Table 6.1 for details).

It is well established that devices at high temperatures will emit infra-red (IR) radiation (TEMPCO Electric Heater 2008). The emission wavelength is function of emissivity and temperature of the emitting body. Hence emitted wavelength (IR) carries the signature of temperature of the heated body (hot-spot). Through appropriate calibrations, the emitted wave the temperature of the hotspot can be estimated.

The emission profile of a micro-heater at different temperatures (873–1473 K) is illustrated (Fig. 6.9). We observe that the emission band is from 0.5 to 10 μm. The peak emission is detected at ~3 μm (0.41 eV). The peak shifts towards shorter wavelengths (higher energy) with increase in temperature. In order to detect such emission spectra by using Silicon based Schottky diode, the barrier height needs to be tailored through appropriate choice of metals. The results are summarized in Table 6.2.

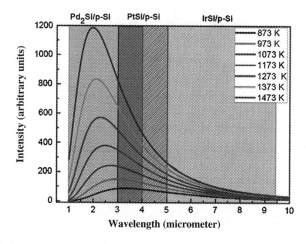

Fig. 6.9 Emission profile of micro-heater, response profile of micro-thermal sensor (multiple technologies)

Table 6.2 Barrier height for different Schottky metal-silicon combinations

Metal contact	Barrier height (eV) with n-Si	λ_c (μm)	Barrier height (eV) with p-Si	λ_c (μm)
PtSi	0.87	1.47	0.24	5.6
IrSi	0.97	0.27	0.13	9.5
Pd$_2$Si	0.76	1.63	0.34	3.7
TiSi$_2$	0.58	2.13	0.53	2.3
WSi$_2$	0.67	1.85	0.44	2.8
CoSi$_2$	0.67	1.90	0.44	2.9
NiSi$_2$	0.66	1.88	0.45	3.26
MoSi$_2$	0.64	1.93	0.47	2.6
TaSi$_2$	0.59	2.10	0.52	2.38

It is observed that the Schottky diode detects photons with energy equal to or greater than the cut-off wavelength (lower energy) and shows highest sensitivity at the cut-off wavelength. In addition, proper selection of semiconductor material is required depending on the temperature band. For high temperatures n-type silicon is preferable whereas for low temperatures p-type silicon is a more appropriate choice.

In addition to metal-silicon combinations noted in Table 6.2 we have also explored metal silicide–silicon based IR photodetectors. Specifically, Palladium Silicide (Pd$_2$Si) on p-type silicon (p-Si) exhibits good response in the 1–3 μm band. Next, Platinum Silicide (PtSi) on p-Si is operational in the 3–5 μm band. Finally, Iridium Silicide (IrSi) on p-Si is suitable for the 5–10 μm band. The application is as a Focal Plane Array (FPA) for commercial infrared imaging systems (Kosonocky et al. 1985; Konuma et al. 1992).

6.4.1 Photo-Detector Fabrication Process Flow

The process flow for Aluminum—n-type Silicon based Schottky diodes noted below. The process flow for PdSi, PtSi and IrSi on p-type Silicon will be similar. All processes are standard CMOS compatible process flow as noted below. Both n-type (phosphorus doped) and p-type (boron doped) silicon (100) wafer of resistivity 0.2–0.4 Ω-cm have been utilized.

1. RCA cleaning was performed on 50 mm diameter silicon wafer.
2. Then the wafer was patterned by lithographic technique by using iron oxide photo-mask. The Double Side Aligner (DSA) in use has a typical best case resolution of 1 μm. The feature dimensions on the top side of silicon substrate was defined and inspected by microscope which was found to be of the order of 5 μm and spacing of 10 μm.
3. Native oxide (SiO_2, few mono-layers thick) removed by dipping wafer in Buffered Hydrofluoric (BHF) etchant solution (typically a 10:1 mixture of 40 % ammonium fluoride, NH_4F and HF) for 15 s. This native oxide is formed during the process due to ambient humidity and oxygen.
4. Subsequently, the wafer was then loaded on to the high vacuum, thermal evaporator system to deposit aluminum (Al) nano-scale films. 30 nm thick Al was deposited by thermal evaporation technique. Deposition was performed in base vacuum of $5e^{-6}$ mbar and pressure of $5e^{-3}$ mbar.
5. Metal film patterns (micron scale) achieved using acetone solvent for lift-off of underlying photoresist.
6. Metal film subjected to rapid thermal annealing at 573 K for 30 s to enable formation of metal-semiconductor (Schottky) junction.
7. Finally back-side ohmic contact was realized by oxide removal, metal deposition and annealing.

The fabrication process steps have been illustrated below in Fig. 6.10 (Roy et al. 2013; Roy et al. 2014). Additional details about micro-fabrication process steps are available at this reference (Campbell 2001). The next step is testing.

6.4.2 Photo-Detector Calibration and Test Setup

A broad-band tunable optical source was used such that the emission wavelength could be varied in the infra-red. The system was calibrated by standard germanium detector and the target diode was kept under DC bias by external unit. The measurement was performed by applying different reverse bias while response was measured in terms of Amp/watt-nm. Currently testing with micro-heater and photo-detector is in progress (Fig. 6.11).

The logarithmic plot of Current versus Voltage indicates the Schottky nature of diode characteristics (Fig. 6.12). The zero bias current was observed to be 100 nA. The linear plot of I–V characteristics is also shown (Fig. 6.13).

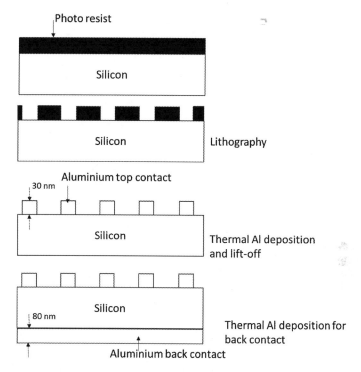

Fig. 6.10 Process flow for aluminum and n-type silicon Schottky diode

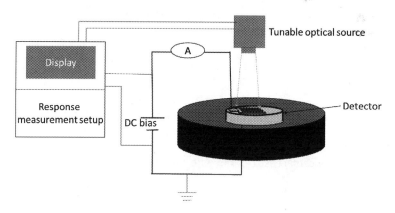

Fig. 6.11 Photo-detector response measurement system

The key challenge is to calibrate photo-detector such that it is able to perform as a thermal sensor. While referring to micro-heater emission characteristics (Fig. 6.9) it can be observed that the source intensity increases with temperature (for same wavelength). Measurement at a particular wavelength will help improve

Fig. 6.12 I–V characteristics (log plot) for aluminum and n-type Schottky diode

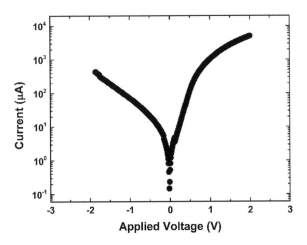

Fig. 6.13 I–V characteristics (linear) for aluminum and n-type Schottky diode

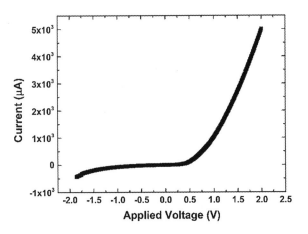

the accuracy of temperature measurements. This will require design and development of gratings that act as a filter. We are exploring the possibility that such gratings can be tunable as well.

A second advancement is possible with the use of two photo-detectors with a distinct yet overlapping IR response (Fig. 6.9).

6.5 Avalanche Photo Diodes (APD) for Thermal Sensing

With the advent of CMOS technology, band-gap based thermal sensors (Hanet al. 2008; Miribel-Catala et al. 2001) have been developed and have certain advantages (for example, ability to integrate in test structures). The temperature sensing range is typically 223–423 K (Texas Instruments 2013) with an accuracy of

Fig. 6.14 Avalanche photo diode (APD) with P+–I–P–N+ dopant profile

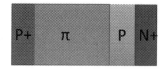

0.5 K. These devices utilize the dependence of some electrical property of device (for example, base to emitter voltage or V_{BE}) on temperature.

Recently, Avalanche Photo Diodes (APDs) have been investigated for their ability to detect very low flux of photons (extending to even single photon detection). The outstanding characteristics of APDs include high gain, high quantum efficiency and low noise. APDs are operated in reverse bias and proximate to the breakdown voltage region of the device.

The APD structure contains two regions: photon absorption region and impact ionization region. Photon absorption region is the depleted π region (Fig. 6.14). Photon absorption results in generation of electron hole pairs. The charge carriers are drifted by the electric field in the π region. Holes are drifted towards the P+ region. Electrons are drifted towards the P–N+ junction. A very high electric field is set up at the P–N+ junction due to the reverse bias. The accelerated electron due to impact ionization generates more charge carriers at the junction. The generated carriers in turn impact ionize and generate more carriers and thereby carrier amplification occurs.

Studies have shown that the breakdown voltage of the diode depends very strongly on temperature. The reported temperature co-efficient for break-down voltage is ~50 mV/K for a silicon based APD. This is a considerable disadvantage (when considering normal APD operation as a photo-sensor) and affects the functionality of APDs due to variations in gain in response to temperature change. Accordingly, research groups have been focused on developing compensation schemes for temperature invariant circuit design (Liet al. 2012). Instead we propose to exploit the temperature dependence of break-down voltage V_{BD} (and how it is affected by device architecture and process technology) for the application of APD as an on-chip, proximate (as opposed to remote) temperature sensor.

Simulations have been performed in order to determine the temperature co-efficient of the breakdown voltage of an avalanche photo diode. Based on these results it is conceivable to use an APD as a micro temperature sensor which can be easily integrated on a silicon chip for proximate thermal sensing with a high level of accuracy.

The breakdown voltage of a diode is theoretically defined as the reverse bias voltage at which the diode current magnitude abruptly changes from zero to infinity (Pierret 1996). For practical purposes, the diode current increases by few orders of magnitude (for example from pico-Amp to micro-Amp). The breakdown voltage can be experimentally measured by incrementally increasing the reverse voltage and recording the voltage at which diode current crosses a predefined value (1 μ-A) (Pierret 1996). This voltage is termed as breakdown voltage (V_{BD}) of the diode. The cross-sectional schematic of silicon (P+–P–N) APD with trench isolation is shown (Fig. 6.15). Simulated temperature dependent reverse I–V characteristics of APD (Fig. 6.16) are in reasonable agreement with experimentally reported data.

Fig. 6.15 Cross-sectional
schematic of silicon
avalanche photo diode (APD)

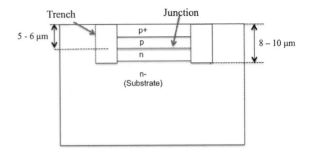

Fig. 6.16 I–V characteristics
(simulated) of silicon
avalanche photo diode (APD)
(Temp in K)

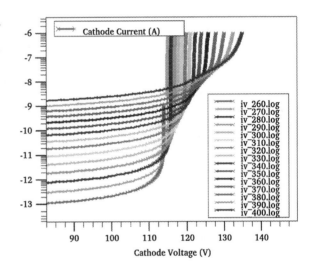

It can be observed that the breakdown voltage V_{BD} of the APD increases as the temperature of the device is increased. The explanation for this phenomenon is as follows. When the charge carriers in the APD are subjected to an increase in temperature, they absorb heat while being transported across the avalanche region under the influence of high electric field and lose part of their energy to the crystal lattice as phonons. Accordingly, the mean free path (λ) decreases. As a result the ionization rate and multiplication factor decreases with increase in temperature. For a specific depletion region width (w), the breakdown voltage (V_{BD}) increases with temperature (Su et al. 1979).

6.5.1 TCAD Simulation Studies

The simulated characteristics for the silicon based APD (Fig. 6.15) has been performed using the Silvaco TCAD platform. APD structures with different V_{BD} corresponding to different doping concentrations have been studied in order to understand the effect on the following parameters:

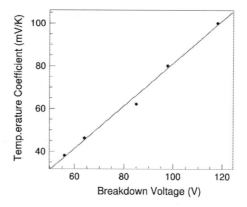

Fig. 6.17 Temperature coefficient (mV/K) of silicon avalanche photo diode (APD) as a function of break-down voltage (simulated results)

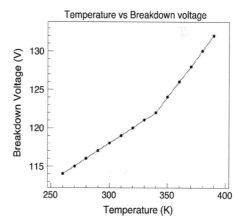

Fig. 6.18 Variation of breakdown voltage with temperature ($V_{BD} = 118$ V)

1. Temperature co-efficient of breakdown voltage V_{BD}
2. Variation in the *temperature coefficient* as a function of V_{BD} of device

The temperature coefficient of the breakdown voltage V_{BD} is observed to be higher in devices with higher breakdown voltage (Fig. 6.17). Diode structures with low doping concentrations (larger depletion width, w) will cause the temperature coefficient to increase. Furthermore, breakdown voltage dependence on technology can be studied following the report that V_{BD} increases with band-gap (Su et al. 1979). Therefore, Ge (0.66 eV @ 300 K) based APDs will have the lower breakdown voltage (V_{BD}) as compared to Si (1.11 eV) and GaAs (1.43 eV).

The variation of temperature coefficient as function of breakdown voltage is shown in Fig. 6.17. Avalanche Photo Diodes with higher breakdown voltage will exhibit enhanced sensitivity. This feature is limited in its usefulness as it would be difficult to realize (and operate) on-chip APDs with very large breakdown voltages. The target range is 100–150 V.

The variation of V_{BD} with temperature for a specific APD is simulated (V_{BD} = 118 V @ 300 K) and the characteristics are shown (Fig. 6.18). It can be observed that in the lower temperature range (<350 K) the sensitivity value is ~100 mV/K while at higher temperatures (>350 K) it is ~200 mV/K. These values will further depend on device architecture and technology.

6.5.2 APD Fabrication Process Flow

The process flow for an Aluminum–N+ Silicon–P Silicon–P+ Silicon based APD (Fig. 6.19) utilizes high-resistivity (>100-cm), double-side polished, p-type silicon (100) wafer.

1. RCA cleaning step for removal of surface contaminants.
2. Thermal oxidation (SiO_2) as protective mask layer (300–500 nm).
3. Patterning (mask level #1: P-well) by photo-lithographic technique by using iron oxide photo-mask. The Double Side Aligner (DSA) in use has a typical best case resolution of 1 μm.
4. Reactive Ion etching (RIE) of oxide in open windows (SiO_2). While a $CF_4–H_2$ process is preferred, SF_6 can be used as a CF_4 substitute.
5. Boron implantation for creation of P-well region.
6. Oxidation step for proceeding to level #2: N+.
7–8. Lithography (mask level #2) and Oxide etching.
9. Phosphorus implantation for creation of N+ window.
10. Backside etching of oxide.
11. Backside boron implant for P+ window.
12. Oxidation step for proceeding to level #3: Al metal contact.
13–14. Lithography (mask level #3) and Oxide etching (front-side).
15. Aluminum contact deposition (front-side) through DC sputtering.
16. Chemical Mechanical Planarization (CMP) of Aluminum layer (front-side). This step is optional.
17–18. Oxide etch (back-side) and sputter deposition for Aluminum back contact.

Additional details about the specifics of micro-fabrication process steps (oxidation, photo-lithography, reactive ion etch, ion implantation, chemical mechanical planarization) are available at this reference (Campbell 2001).

In conclusion, the breakdown voltage of the APD depends critically on temperature and thus an APD can be utilized for micro thermal sensor application. In future work one can focus on how device sensitivity is dependent on device architectures (specific arrangement of epitaxial grown thin-film layers) and process technology (silicon doping, alloying).

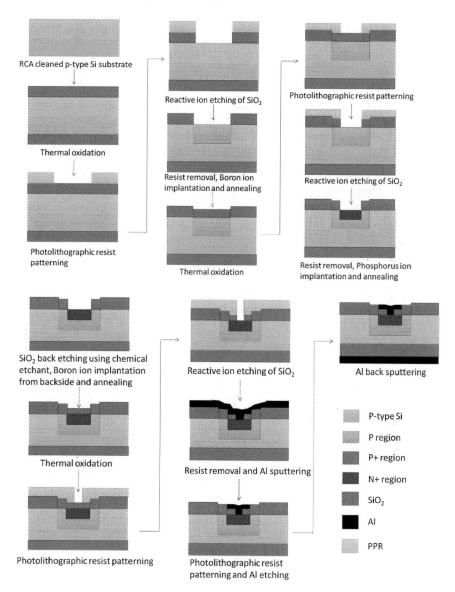

Fig. 6.19 CMOS compatible process flow for silicon avalanche photodiode (APD)

References

Baschuk JJ, Li X (2001) Carbon monoxide poisoning of proton exchange membrane fuel cells. International Journal of Energy Research 25(8): 695–713

Campbell SA (2001) The Science and Engineering of Microelectronic Fabrication, Oxford University Press (2nd edition), USA. ISBN: 0195136055, 9780195136050

Das A, Chatterjee U, Duttagupta SP, Gandhi MN (2014) Study of Gamma Irradiation Effect on Sol-Gel derived Lithium Borate Glassy Film based Metal Insulator Semiconductor (MIS) Structure. Journal of Electron Device (accepted for publication)

Das A, Duttagupta SP, Gandhi MN (2012) TCAD and Compact Models for SiCSchottky Diode as Neutron Induced Ion Detector.Paper presented at the *International Conference on Electronics, Communication and Signal Processing,* Nagpur, India.

Dimble SS (2012) Fabrication, packaging and testing of PEM fuel cells.Dissertation, Indian Institute of Technology Bombay, Mumbai, India

Dimble SS, Ramesh P, Duttagupta SP (2011) Effect of segmented contacts on fuel cell performance using 3D modeling. Paper presented at the ASME International Conference on Energy and Electrical Systems, Malaysia

Duan D (2012) Sensing, Communications and Monitoring for the Smart Grid, Colorado State University Fort Collins, Colorado, USA

Dziedzic A, Golonka, L, Hrovat M, Kita, J, Belavic D (2005) LTCC resistors and resistive temperature sensors-chosen electrical and stability properties. Paper presented at the 28th International Spring Seminar on Electronics Technology (ISSE 2005), Wiener Neustadt, Austria

Golonka LJ (2006)Technology and applications of low temperature co-fired ceramic (LTCC) based sensors and microsystems. Bulletin of the Polish Academy of Sciences, Technical Sciences 54(2):221–231

Gongora-Rubio MR, Espinoza-Vallejos P, Sola-Laguna L, Santiago-Aviles JJ (2001) Overview of low temperature co-fired ceramic tape technology for meso-system technology. Sensors and Actuators A 89:222–241

Gregory OJ, You T (2005) Ceramic temperature sensors for harsh environments. IEEE Sensors Journal 5(5):833–838

Han DO, Kwon YI, Park TJ, Park HC (2008) A CMOS temperature sensor with calibration sensing technology. Proceedings of the 3^{rd}International Conference on Sensing Technology (ICST 2008): 496–499

Heng G, Black WZ (1986) Thermal modeling for microelectronic chips packages. Technical Report, Georgia Institute of Technology.https://smartech.gatech.edu/bitstream/handle/1853/3 8388/e-25-669_306533_fr.pdf. Accessed 16 June 2014

Hrovat M, Belavic D, Kita J, Holc J, Cilensek J, Golonka L, Dziedzic A (2007) Thick-film PTC thermistors and LTCC structures: The dependence of the electrical and microstructural characteristics on the firing temperature. Journal of the European Ceramic Society 27(5):2237–2243

Huang K, Goodenough JB (2009).Solid Oxide Fuel Cell Technology: Principles, Performance and Operations. London: Woodhead Publishing.

Jain YK, Khanna VK (2007) Thick Film, LTCC or silicon microhotplate for gas sensor and other applications. Paper presented at the 14^{th}International Workshop on Physics of Semiconductor Devices (IWPSD 2007), IIT Bombay

Jurków D, Malecha K, Golonka LJ (2008) Investigation of LTCC thermistor properties. Materiał yElektroniczne36(4):133–138

Kita J, Dziedzic A, Golonka LJ, Bochenek A (2000) Properties of laser cut LTCC heaters. Microelectronics Reliability 40:1005–1010

Konuma K, Tohyama ST, Tanabe A, Teranishi N, Masubuchi K, Saito T and Muramatsu T (1992) A standard-television compatible 648×487 pixel Schottky-barrier infrared CCD image sensor. IEEE Transactions on Electron Devices 39(7): 1633–1637

Kosonocky WF, Shallcross FV and Villani TS (1985) 160×244 element PtSiSchottky-barrier IR-CCD image sensor. IEEE Transactions on Electron Devices 32(8): 1564–1573

Kulawik J, Szwagierczak D, Witek K, Skwarek A, Gröger B (2012) Multilayer perovskite based thermistors fabricated by LTCC technology. Paper presented at the 2^{nd}International Applied Physics & Materials Science and Engineering Conference (APMAS 2012), Antalya, Turkey

Kulkarni SG (2014) Investigations on electrodes and electrolyte materials for the fabrication of integrated low temperature solid oxide fuel cells in low temperature co-fired ceramic structures. Dissertation (submitted), University of Pune, Pune, India

Kulkarni SG, Ramesh P, Duttagupta SP,Phatak GJ (2012) Nanocrystalline gadolinium doped ceria (Ce0.8Gd0.2O3-δ) for oxygen sensor and solid oxide fuel cell applications. Paper presented at the1st International Symposium on Physics and Technology of Sensors (ISPTS 2012), Pune, India

Lavenuta G (1997) Negative temperature coefficient thermistors Part(I): characteristics, materials, and configurations. http://www.sensorsmag.com/sensors/temperature/negative-temperature-coefficient-thermistors-part-i-characte-811. Accessed 16 June 2014

Li Z, Xu Y, Liu C, Li Y, Li M, Chen Y, Wang Y, Lu B, Cui W, Huo J, Chen T, Han D, Hu W, Li C, Li W, Liu X, Wang J, Yang Y, Zhang Y, Zhu Y, Li G, Zhao J, Wang J, Pu N, Li X (2012) A gain control and stabilization technique for silicon photo-multipliers in low-light-level applications around room temperature. Nuclear Instruments and Methods in Physics Research A 695: 222–225

Lin PC, Park BY, Madou MJ (2008) Development and characterization of a miniature PEM fuel cell stack with carbon bipolar plates, Journal of Power Sources 176(1): 207–214

Liu F, Yi B, Xing D, Yu J, Zhang H (2003) Nafion/PTFE composite membranes for fuel cell applications. Journal of Membrane Science 212: 213–223

Madou MJ, Park BY (2006) Design, fabrication, and initial testing of a miniature PEM fuel cell with micro-scale pyrolyzed carbon fluidic plates. Journal of Power Sources 162: 369–379

Miribel-Catala P, Montane E, Bota SA, Puig-Vidal M, Samitier J (2001) MOSFET-based temperature sensor for standard BCD smart power technology. Microelectronics Journal 32(10–11): 869-873

Mitra S (2014) Sensor array based performance threat estimation for operational safety of PEMFC based energy generation system. Dissertation, Indian Institute of Technology Bombay, Mumbai, India

Mitra S, Ramesh P, DuttaguptaSP (2013) Localized O2 starvation threat estimation and rapid O2 compensation technique in air PEMFC. 1stInternational Conference on Power Engineering, Energy and Electrical Drives (PEED 2013), Cambridge, MA, USA

Mitra S, Ramesh P, Bhattacharyya M, DuttaguptaSP (2012) Multimode sensing technique for carbon monoxide plume tracking and forecasting for reliable field deployed air breathing PEM fuel cell operation.Paper presented at the 1st International Symposium on Physics and Technology of Sensors (ISPTS 2012), Pune, India

MolDovan C, Nedelcu O, Johander P, Goenaga I, Gomez D, Petkov P, Kaufmann U, Ritzhauptkleissl HJ, Dorey R, Peterson K(2007)Ceramic micro heater technology for gas sensor. Romanian Journal of Information Science and Technology 10:43–52

Nowak D, Mis E, DziedzicA(2009)Fabrication and electrical properties of laser-shaped thick film and LTCC microresistors. Microelectronics Reliability 49:600–606

Okyay AK, Nayfeh AN, Saraswat KC, Yonehara T, Marshall A, McIntyre PC (2006) High-efficiency metal–semiconductor–metal photodetectors on heteroepitaxially grown Ge on Si. Optics Letters 31(17): 2565–2567

Patil TC, Duttagupta SP (2014) Hybrid self–sustainable green power generation system for powering green data center.Paper presented at the International Conference on Control, Instrumentation, Energy and Communication (CIEC 2014), Kolkata, India

Patil TC, Kulkarni SG, Duttagupta SP, Phatak GJ (2013) Oxygen ion transport through the electrolyte in solid oxide fuel cell. Paper presented at the International Conference on Renewable Energy Research and Applications (ICRERA) 2013, Madrid, Spain

Patnaik B, Aswani U, Sarkar G, Duttagupta SP (2014) Image aided dynamic reconfiguration of SPV array under non-uniform illumination. Paper presented at the 40th IEEE Photo-Voltaics Specialists Conference (IEEE-PVSC), Denver, Colorado, USA

Patnaik B, Mohod JD, Duttagupta SP (2012) Dynamic loss comparison between fixed-state and reconfigurable solar photovoltaic array.Paper presented at the 38th IEEE Photo-Voltaics Specialists Conference (IEEE-PVSC), Austin, Texas, USA

Patnaik B, Sharma S, Trimmurthulu E, Duttagupta SP, Agarwal V (2011)Reconfiguration strategy for optimization of solar photovoltaics array under Non-Uniform Illumination Conditions. Paper presented at the 37th IEEE Photo-Voltaics Specialists Conference (IEEE-PVSC), Seattle, Washington, USA

Pierret RF (1996) Semiconductor Device Fundamentals. Addison-Wesley (Pearson).International Edition; ISBN13: 9780131784598, ISBN10: 0131784595

Pisarkiewicz T, Sutor A, Potempa P, Maziarz W, Thust H, Thelemann T(2003) Micosensor based on low temperature cofired ceramics and gas sensitive thin film. Thin Solid Films 436:84–89

Ramesh P (2014)Design and development of field-testable micro PEM fuel cells. Dissertation (submitted), Indian Institute of Technology Bombay, Mumbai, India

Ramesh P, Duttagupta SP (2014) Three dimensional model analysis of a PEM fuel cell with ceramic flow field plates. International Journal of Electrochemical Science, 7(8): 4331–4344

Ramesh P, Duttagupta SP (2013) Study of the effect of dimensions on micro fuel cell performance using 3-Dmodeling. International Journal of Renewable Energy Research, 3(2): 353–358

Ramesh P, Duttagupta SP (2012) Distributed array of multi sensor nodes for dynamic detection of smoke plume,Paper presented at the International Conference on Electronics, Communication and Signal processing, Nagpur, India

Ramesh P, Dimble SS, Duttagupta SP (2011) Performance of miniature fuel cells with segmented contacts attached to the GDL, Paper presented at the 2[nd]IEEE International Conference on Electric Power Systems, Sharjah

Rogalski A (2003) Infrared detectors: status and trends. Progress in Quantum Electronics 27: 59–210

Roy S, Midya K, Duttagupta SP, Ramakrishnan D (2014) Nano-scale NiSi and n-type Silicon based Schottky barrier diode as a NIR detector for room temperature operation. Journal of Applied Physics (accepted for publication)

Saddow S, Agarwal A (2004) Advances in Silicon Carbide Processing and Applications. Norwood (MA 02062 USA): Artech House

Schwarz C, Kanel HV (1996) Tunable infrared detector with epitaxial silicide/silicon heterostructures. Journal Applied Physics 79(11): 8796–8807

Steinberg DS (1980)Cooling Techniques for Electronic Equipment. New York: John Wiley & Sons

SuYK, ChangCY, Wu TS, Liu BD (1979) Temperature-dependent characteristics of a reach-through avalanche photodiode (RAPD)in Ge, Si and GaAs. Optical and Quantum Electronics 11(5): 377–384

TEMPCO (2008) Ceramic Infrared E-Mitters http://www.tempco.com/Infrared/CeramicEmitterHub.html. Accessed 16 June 2014

Texas Instruments (2013) LM35 Precision Centigrade Temperature Sensor. http://www.ti.com/lit/ds/symlink/lm35.pdf. Accessed 16 June 2014

Vasudev A, Kaushik A, Jones K, Bhansali S (2013)Prospects of low temperature co-fired ceramic (LTCC) based microfluidic systems for point-of-care biosensing and environmental sensing. Microfluidics and Nanofluidics 14:683–702

Wang CY, Wang Y (2006) A non isothermal two phase model for polymer electrolyte fuel cells. Journal of the Electrochemical Society. 153(6): A1193–A1200

Zhang S, Yuan X, Wang H, Merida W, Zhu H, Shen J, Wu S, and Zhang J (2009) A review of accelerated stress tests of MEA durability in PEM fuel cells, International Journal of Hydrogen Energy. 34(1):388–404

Zhao T (2009) Micro Fuel Cells: Principles and Applications. London: Academic Press

Author's Biographies

Choobineh, Leila Leila Choobineh is a graduate student in the Microscale Thermophysics Laboratory at the University of Texas at Arlington, TX. Her research interests include thermal engineering of three-dimensional integrated circuits, microfabrication, etc. She completed her Ph.D. in 2014, and prior to that, received her M.S. from Shahrekord University and B.S. from Shiraz University, both in Mechanical Engineering.

Duttagupta, SP Siddhartha Prakash Duttagupta received a Bachelor of Technology (Honours) in Electronics and Communication Engineering from the Indian Institute of Technology, Kharagpur, India and an MS and a Ph.D. in Electrical and Computer Engineering from the University of Rochester, Rochester New York, USA. Dr. Duttagupta has previously held the position of Assistant Professor in Electrical and Computer Engineering, Boise State University, Boise, Idaho, USA, and Director of Technology, Solar Integrated Technologies, USA. He is presently an Associate Professor in the Department of Electrical Engineering at the Indian Institute of Technology, Bombay, Mumbai, India. Dr. Duttagupta's research interests are in the areas of (i) Micro/Nano Sensor Technology Optimization and Application, and (ii) Sensor Integrated Electronic Circuits and Systems-Design, Development and Field Deployment. The significance of research contributions following Google Scholar is: total citations 2533, h-index 19, i10-index 28. He has published and presented more than 100 papers in Journals and International Conferences, and has acted as a referee for leading international journals. He has authored books on Ambulatory ECG (Ambulation Analysis in Wearable ECG) and Electronics Cooling (Thermal Management of Electronic Circuits) and has filed for 15 Indian and US Patents in the general area of Micro/Nano Technology and Electronic Systems design and development. Dr. Duttagupta has participated as an investigator in multiple sponsored research projects from funding agencies based in India (Department of Science and Technology) and in the USA (National Science Foundation). He has also been involved in sponsored, collaborative research and development with organizations such as Crompton Greaves, Datar Power Management, Nipro Glass and Tube, and Tata Consultancy Services.

© Springer Science+Business Media New York 2015
C.M. Jha (ed.), *Thermal Sensors*, DOI 10.1007/978-1-4939-2581-0

Huynh, Thu Thu Huynh is a thermal analyst at Intel Corporation. She is responsible for system and component thermal analysis and reference thermal solution design in the Datacenter and Connected Systems Group at Intel Corporation. She received the B.S degree in Chemical Engineering from Northwestern University, Evanston, IL, USA in 2008. She began her career as a Rotational Engineer at Intel Corporation, with rotations in research, assembly/test and product development in the Rotational Engineering Program.

Jain, Ankur Ankur Jain is an Assistant Professor in the Mechanical and Aerospace Engineering Department at the University of Texas, Arlington where he directs the Microscale Thermophysics Laboratory. His research interests include energy conversion devices, microscale thermal transport, bioheat transfer, etc. He previously held research and development positions in leading semiconductor companies including AMD and Freescale Semiconductor, and worked at Molecular Imprints Inc., a nanotechnology-based startup company. He received his Ph.D. (2007) and M.S. (2003) in Mechanical Engineering from Stanford University, and his B.Tech. (2001) in Mechanical Engineering from the Indian Institute of Technology (IIT), Delhi with top honors. His research has been supported by National Science Foundation (NSF), Office of Naval Research (ONR), Department of Energy (DoE) and Indo-US Science and Technology Forum (IUSSTF).

Jha, Chandra Mohan (CM) CM Jha is a Staff Thermal Engineer in Intel Corporation (Sept 2008—present). He is responsible for research and development of advanced thermal management technologies and thermal reliabilities for Intel microprocessor packages. He received the M.S. and Ph.D. degrees in mechanical engineering from Stanford University (2004–2008) and worked on MEMS thermal isolation project (sponsored by DARPA), Vibration Energy Scavenging (AUDI project) and Shape Optimization (Draper Lab project). His work in Bhabha Atomic Research Centre, Mumbai, India (1997–2004) helped the organization become self-reliant in certain mechanical equipment and systems which had to be earlier imported, thereby saving huge foreign exchange of India government. He received the B.E. degree in mechanical engineering from Birla Institute of Technology, Mesra, India, in 1996. He has published more than 30 peer reviewed journal/conference publications with over 250 citations and holds multiple patents. He has reviewed several papers for ASME and IEEE conferences and journals, and is currently a co-editor of the design and analysis section of the Intel IATTJ Journal (Intel Assembly and Test Technology Journal).

Krishnan, Gopi Gopi Krishnan is a Senior process Engineer at Intel. In this role he is part of the team that develops the Integrated Heat Spreader assembly process for Intel's next generation products, from technology development through high volume manufacturing ramp. He holds a Ph.D. in mechanical engineering from the University of Colorado-Boulder with an emphasis in thermal/fluid science. Following graduation he was a post-doctoral researcher at the National Renewable Energy Laboratory in Golden, CO., focusing on the characterization of air based thermal management methods for power electronics in electric vehicles. His other interests lie in resource conservation in manufacturing and driving employee/community driven environmental sustainability initiatives.

Kulkarni, SG Shrikant Gajanan Kulkarni received a Master of Science in Physics from Shivaji University, Kolhapur and is presently pursuing a doctorate at the University of Pune. Presently, he is a Senior Research Fellow in the Centre for Materials for Electronics Technology (C-MET), Pune. Since 2006 he is working in Electronic Packaging group on different research positions. His field of study is Low Temperature Co-fired Ceramics (LTCC) based micro-fluidic devices and high temperature fuel cells.

Phatak, GJ Girish Jayant Phatak received a Master of Science in Physics from University of Pune, Pune, India and a Ph.D. degree from Electrical Engineering Department, Indian Institute of Technology, Bombay, Mumbai, India. He was a faculty with Electronic Science Department, University of Pune and later joined as Scientist at Centre for Materials for Electronics Technology (C-MET), Pune. Dr. Phatak received MONBUSHO fellowship from Govt. Japan and worked for his Post Doctoral Research at Nagoya University, Nagoya, Japan from 1997 to 1999. After his initial work in the area of thin films and Si based devices, he shifted to the area of Hybrid Micro Circuit (HMC) and Surface Mount Technology (SMT) materials, popularly known as Thick film materials. He has contributed to the development of paste families of Ag and Ag-Pd based conductor pastes, Birox based resistor pastes and solder pastes. The Solder paste technology has since been has been transferred to the Industry. Recently, Dr. Phatak has taken a major initiative in the area of Electronic Packaging, and has set up a facility for the fabrication of Low Temperature Co-fired Ceramic (LTCC) based ceramic circuits and packages, with help of generous support from the National Programme on Smart Materials (NPSM). He is currently involved in several projects of national importance where he is contributing in development of special packages and circuits. He has above 32 research papers to his credits in renowned Journals as well as several more contributions to International and National conferences.

Pushpagandan, Ramesh Ramesh Pushpagandan received a Bachelor of Engineering from Cochin University of Science and Technology, Kerala, India with Electronics Engineering as specialization and a Master of Engineering in Instrumentation Engineering from Kerala University, Kerala, India. Presently he is pursuing a doctorate at the Indian Institute of Technology Bombay, Mumbai, India involving the design, fabrication and testing of micro fuel cells. He is an Assistant Professor in the Department of Electronics and Communication Engineering, College of Engineering Munnar, Kerala, India. He has authored 5 international journal publications and is an active IEEE member for the past 16 years and has held various positions in IEEE India Council. Also he is a member of Indian Society for Technical Education and Institute of Engineers India. His areas of interest are renewable energy, micro fuel cells, instrumentation and VLSI Design.

Roy, S Sandipta Roy received a Bachelor of Science and Master of Science in Physics from University of Kalyani, West Bengal, India and a Master of Technology in Materials Science from the Indian Institute of Technology Kharagpur, Kharagpur, India. Presently he is pursuing a doctorate at the Indian Institute of Technology Bombay, Mumbai, India involving the design, fabrication and testing of Infra-Red detectors for the purpose of toxic gas monitoring.

Sanchez, Jaime Jaime Sanchez has a Ph.D. in Mechanical Engineering from the University of Kentucky and is a licensed Professional Engineer. He has written various archival publications in the areas of nanoscale heat transfer and molecular simulations, as well as conference presentations on test challenges of high volume manufacturing of microprocessors. He joined Intel in 2008 and his team is in charge of research and development of test equipment used in high volume manufacturing. His work is focused on thermal-mechanical design and challenges for test to support all products in Intel's roadmap.

Shukla, RA Raghunandan Atul Shukla received a Bachelor of Engineering with specialization in Electronics and Telecommunications from University of Pune. Presently he is pursuing a doctorate at the Indian Institute of Technology Bombay, Mumbai, India involving the design, fabrication and testing of Avalanche Photo Detectors with capability for single photon detection. He is a Scientific Officer (C) with the Tata Institute of Fundamental Research, Mumbai, India.

Printed in the United States
By Bookmasters